ASTOUNDING AVERAGES!

ASTOUNDING

90 Acres of Pizza a Day

...and more!

AVERAGES!

Dean D. Dauphinais and
Kathleen Droste

VISIBLE INK PRESS

Detroit

New York

Washington, D.C.

Toronto

ASTOUNDING AVERAGES!

90 Acres of Pizza a Day
... and more!

Copyright © 1995 by Visible Ink Press

Most Visible Ink Press™ books are available at special quantity discounts when purchased in bulk by corporations, organizations, or groups. Customized printings, special imprints, messages, and excerpts can be produced to meet your needs. For more information, contact Special Markets Manager, Visible Ink Press, 835 Penobscot Bldg., Detroit, MI 48226. Or call 1-800-776-6265.

Published by Visible Ink Press™
a division of Gale Research Inc.

Visible Ink Press is a trademark of Gale Research Inc.

Art Director: Mary Krzewinski

Library of Congress Cataloging-in-Publication Data
Dauphinais, Dean D., 1961-
  Astounding averages! : 90 acres of pizza a day ... and more!/
Dean D. Dauphinais and Kathleen Droste.
    p. cm.
  Includes index.
  ISBN 0-7876-0695-2
  1. Statistics. I. Droste, Kathleen D.   II. Title.
HA155.D38   1995
519.5—dc20   95-34040
                                                        CIP

519.5
DAU
1995

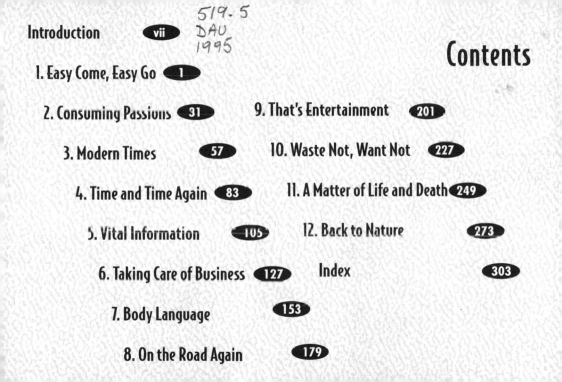

Introduction

*H*i there! Welcome to *Astounding Averages! 90 Acres of Pizza a Day . . . and more!* Be prepared: this book is chock-full of fascinating information taken from the *Gale Book of Averages.* It contains hundreds of the most intriguing, mind-boggling, and—you guessed it—astounding averages you'll ever want to know. Like how many calories the average kiss burns up and how much pizza Americans consume each day. (C'mon. Admit it. Weren't you just *dying* to know where the subtitle of this book came from?) So kick off your shoes, pull up a footstool, and dive into this little gold mine. By the way, if you're looking to be astounded by information on a specific subject, be sure to check out the **Index** in the back of the book. It's not really astounding, but it *is* average. And you might be, too! But you'll never know until you start reading. . . .

ASTOUNDING AVERAGES!

Easy Come, Easy Go

Man's Best Friend?

*T*he average cost of owning a dog—based on an average life of eleven years—is $13,350; $480 for the initial investment and $1,170 annually.

Who says so? *The New York Times,* "The Cost of That Dog in the Window," June 13, 1992, p. 48. *Who said so first?* American Kennel Club.

EASY COME, EASY GO

Having My Baby

*I*n 1991 the average total childbirth cost in the United States—including maternity care and doctor's and hospital's charges—was $4,720 for a normal delivery and $7,826 for a cesarean section.

Who says so? USA TODAY, "Costs of Having a Baby," March 21, 1993, p. 15A. *Who said so first?* Health Insurance Association of America, *Source Book of Health Insurance Data, 1992.*

The Poop on Diapers

*T*he average per-diaper cost for cloth diapers is 7 to 9 cents; for cloth diapers from a diaper service, 13 to 17 cents; and for disposable diapers, 25 cents.

Who says so? William Rathje and Cullen Murphy, "Cotton vs. Disposables: What's the Damage?" *Garbage,* October/November 1992, pp. 29–30.

5

College Education Not Included

*a*ccording to a government study done by the Agriculture Department's Family Economics Research Group, the average cost of raising a child from birth to age eighteen—in 1992 dollars—was $128,670, not including college.

Who says so? *The Wall Street Journal,* "Just Think of It All As Another Mortgage," June 2, 1993, p. B1.

ASTOUNDING AVERAGES!

Dear Mom and Dad, Please Send More Money

College students spend an average of $135 per month on snack foods, compact discs, movie tickets, and other discretionary items.

Who says so? Eben Shapiro, "The Media Business—Advertising: New Marketing Specialists Tap Collegiate Consumers," *The New York Times,* February 27, 1992, p. D16.

'Tis the Season to Be Spending

*T*he average household spent $400 on holiday gifts in 1992, compared with $375 in 1991. The biggest spenders were those between the ages of forty-five and fifty-four, whose average spending was $467. One in three households spent less than $200.

Who says so? U.S. News & World Report, "A Big Ho Ho Ho for America's Retailers?" December 7, 1992, p. 16.

ASTOUNDING AVERAGES!

What's Left for Santa to Bring?

*a*ccording to Gallup's "North Pole Poll," parents spent an average of $213 per child, ages seven to sixteen years, for Christmas gifts in 1992. This exceeded the $209 children thought their parents should spend.

Who says so? Neely Tucker, "Bottom Line: How Parents Stack Up," *Detroit Free Press*, December 30, 1992, p. 1D.

9

EASY COME, EASY GO

Prom Night

*a*ccording to *Your Prom* magazine, the average cost of a high school prom in 1993 was $1,058: $473 for the boy and $585 for the girl. His total included $20 for flowers, $43 for tickets, $89 for tuxedo rental, and $161 for a limousine. Her total included $14 for a boutonniere, $38 for a visit to the beauty salon, $169 for shoes and accessories, and $200 for a gown.

Who says so? USA WEEKEND, "Prom's Price," May 14–16, 1993, p. 14.

How to Impress a Date

In 1992 the average cost to rent a Lamborghini Diablo was $1,500 per day and 10 cents per mile; a Rolls-Royce Corniche was $750 per day and 50 cents per mile; and a Porsche 911 Turbo was $350 per day and 50 cents per mile.

Who says so? *The New York Times,* "Price Tag: Renting Exotic Automobiles," August 8, 1992, p. 50. *Who said so first?* Budget Rent-A-Car, Car Collection division; Hertz Rent-A-Car; Avis Rent-A-Car; *Motor Trend* magazine; and *Car and Driver* magazine.

11

EASY COME, EASY GO

Here Comes the Bride

*a*ccording to a survey of the readers of *Bride's* magazine, the average cost of a formal wedding in 1990, including the honeymoon, was $19,344. Based on 200 guests and 5 attendants, the total included $82 for the rental of the groom's suit; $963 for the bride's dress and veil; $1,004 for wedding rings; $3,200 for the honeymoon; and $5,900 for the reception.

Who says so? Rachel Powell, "It's One Party Even the Recession Can't Spoil," *The New York Times,* June 23, 1991, p. C10.

Here Comes the Bride (Japanese Style)

The average Japanese Shinto wedding service costs $24,000, but it can easily run as high as $40,000. The average rental fee for a Japanese bridal dress is $1,600; the average cost per wedding guest is $350; and the Japanese custom of serving seven-course meals, with lobster or steak, costs an average of $130 per plate.

Who says so? David J. Morrow, "Japanese Pay for Tradition," *Detroit Free Press*, February 14, 1993, pp. 1Q, 4Q.

Under House Arrest

*I*n 1990 the average monthly mortgage payment was $1,127, or 32.8 percent of household income.

Who says so? U.S. Bureau of the Census, *Statistical Abstract of the United States 1992*. 112th edition. Washington, D.C.: U.S. Department of Commerce, 1992, p. 724. *Who said so first?* Chicago Title Insurance Company, Chicago, IL, *The Guarantor*, bimonthly.

Go South, Young Man

*I*n 1991 the median price of a new home in the South was $100,000; in the Midwest, $110,000; in the West, $142,300; and in the Northeast, $155,400.

Who says so? U.S. Bureau of the Census, *Statistical Abstract of the United States 1992*. 112th edition. Washington, D.C.: U.S. Department of Commerce, 1992, p. 712. *Who said so first?* U.S. Bureau of the Census and U.S. Department of Housing and Urban Development, *Current Construction Reports*, series C25, *Characteristics of New Housing*, annual; and *One-Family Houses Sold*, monthly.

Putting Their Money on the Table

*H*ouseholders under age thirty-five spend an average of $310 per year on furniture, according to a consumer expenditure survey done by the U.S. Bureau of Labor Statistics.

Who says so? Thomas Exter, "Spending Money: Nice Niches for Furniture," *American Demographics*, March 1991, p. 6. *Who said so first?* U.S. Department of Labor, Bureau of Labor Statistics, *Consumer Expenditure Surveys, 1985–1989*.

Spending Power

Match the electric household appliance with its average annual energy cost.

1. Washer/dryer
2. Color TV
3. Range/oven
4. Refrigerator/freezer
5. Microwave oven
6. Dishwasher

A. $ 239
B. $ 77
C. $ 42
D. $ 22
E. $ 12
F. $ 7

(Answers: 1.B 2.D 3.C 4.A 5.F 6.E)

Who says so? U.S. Congress, Office of Technology Assessment, *Changing by Degrees: Steps to Reduce Greenhouse Gases.* Washington, D.C.: February 1991, p. 117

There's No Place Like Home

The average cost of a night at home—consisting of a McDonald's Quarter Pounder, a six-pack of beer, and a new Monopoly game—in Cleveland, Ohio, is $17.57; in Orlando, Florida, $17.97; in Tulsa, Oklahoma, $18.42; in Anchorage, Alaska, $26.69; and in New York City, $27.44. The average for 286 other cities is $18.53.

Who says so? Business Week, "A Burger, a Cold One, and Thou," August 31, 1992, p. 34. *Who said so first?* ACCRA.

ASTOUNDING AVERAGES!

Better Take the Gondola

*I*n 1992 the average price of a gallon of gasoline in the United States was $1.05; in Italy it was $4.71.

Who says so? U.S. News & World Report, "Average Price per Gallon of Gasoline in the U.S.," January 18, 1993, p. 61. *Who said so first?* American Petroleum Institute, Energy Information Administration, OECD.

19

EASY COME, EASY GO

Driving a Hard Bargain

*a*ccording to Dean Witter Reynolds, a New York investment banking firm, in 1992 the average car payment was $308 with a 4½-year loan period. In 1975, the average car payment was $133 with a 3-year loan period.

Who says so? Joann Muller, "Cautious Customers," *Detroit Free Press*, August 24, 1992, pp. 8F–9F.

But It's Nice to Know
It's There If You Need It

*F*rom 1982 to 1992, the average household spent $8,910 on auto insurance but filed only one claim averaging $600.

Who says so? Consumer Reports, "How to Choose the Right Company," August 1992, pp. 489–499.

Footing the Bill

*T*he average household spent $242 on footwear in 1991.

Who says so? U.S. Department of Labor, Bureau of Labor Statistics, *Consumer Expenditure Surveys, 1984–1991.*

ASTOUNDING AVERAGES!

Up in Smoke

The average smoker spends $750 per year on cigarettes.

Who says so? Walecia Konrad and Christopher Power, "Smoking Out the Elusive Smoker," *Business Week,* March 16, 1992, pp. 62–63.

EASY COME, EASY GO

A Taxing Statistic

*I*n 1960 every man, woman, and child paid an average of $629 in federal, state, and local taxes. In 1990 the average was $7,592.

Who says so? Louis Rukeyser, *Louis Rukeyser's Business Almanac.* New York: Simon & Schuster, 1991, p. 166.

ASTOUNDING AVERAGES!

Will That Be Cash or Charge?

The Persian Gulf War cost the United States alone a total of $61 billion. The cost to engage in air combat averaged $175 million per day; the cost to engage in combined air/ground combat averaged $500 million per day.

Who says so? Center for Defense Information, Washington, D.C.

25

You Can Bet on It

*a*mericans wager $90 per capita on lotteries every year. With thirty-two states and the District of Columbia allowing lotteries, $200 billion is bet annually in the United States.

Who says so? Lawrence Dandurand and others, "A Global Framework for Analyzing Gaming Business Units," *Nevada Review of Business & Economics,* Spring–Summer 1991, pp. 2–13.

Take Two Aspirin and
Call Yourself in the Morning

*a*ccording to the American Association of Medical Colleges, the average debt of the 81 percent of medical school graduates who were in debt in 1989 was $42,374, up from $19,700 in 1981. The debt for minority and underprivileged students who completed four years of medical school training was even greater: 91 percent of the graduates were in debt, with an average debt of $48,168; 41 percent had debts exceeding $50,000.

Who says so? William S. Cohen, "Statements on Introduced Bills and Joint Resolutions," *Congressional Record*, January 14, 1991, p. S747.

Let's Talk Turkey

*a*ccording to a survey by the American Farm Bureau, the average cost of a homemade Thanksgiving meal for ten people is $25.95.

Who says so? The American Farm Bureau, Department of Information and Public Statistics, Park Ridge, IL.

The Grand Finale

*a*ccording to a 1989 survey by the Federal Trade Commission, the average funeral-burial package for Americans costs about $3,800. Cremation-based funerals cost an average of $1,500.

Who says so? *American Demographics*, "Changing Styles Bring Cremation Industry to Life," December 1992, p. 25.

EASY COME, EASY GO

The Grand Finale (Japanese Style)

*a*n average funeral service in Tokyo, Japan, costs $20,000, but many run as high as $75,000. This includes mortician and temple fees, flowers, altars, and food and drink for guests. Because of the scarcity of land, most Japanese are cremated, but even a small temple plot can cost $28,000.

Who says so? Karen Lowry Miller, "International Business—Japan: Rest in Peace . . . With Lasers, Smoke, and Synthesizer Music," *Business Week,* September 16, 1991, p. 48.

Consuming Passions

Time to Re-tire

*a*mericans wear out an average of 240 million automobile tires each year. That's almost one tire for every man, woman, and child in the United States.

Who says so? *Consumer Reports,* "Tired Out," April 1991, p. 211.

Changing Time

Every baby who wears disposable diapers uses an average of 4,500 in his or her infancy. In 1991 alone, 17 million disposable diapers were sold in the United States.

Who says so? Michael Specter, "Among the Earth Baby Set, Disposable Diapers are Back," *The New York Times,* October 23, 1992, p. A1.

Smells Like Teen Spirit

*T*eenage boys use deodorant an average of 6.5 times per week; teenage girls use deodorant an average of 7.4 times per week.

Who says so? USA TODAY, "Teens Preen: Boys vs. Girls," May 26, 1993, p. 1D. *Who said so first?* Teenage Research Unlimited.

35

Honey, Can I Borrow Your Razor?

*a*ccording to Gillette Company, women shave an average of 412 square inches of skin, men 48 square inches. But men shave more often and buy more blades than women: an average of 30 per year for men, 10 for women.

Who says so? *The Wall Street Journal,* "Gillette's New Sensor Picks Up a Big Edge in Razors for Women," December 17, 1992, p. B10.

ASTOUNDING AVERAGES!

Brown Baggin' It

*a*ccording to Harry Balzer of NPD Group, a consumer research firm in Chicago, 11 percent of all Americans carry their lunch away from home each day. In 1992 Americans carried an average of 53 meals away from home, of which only 58 percent contained a sandwich. In 1984, the average American carried only 42 meals away from home, but 71 percent contained a sandwich.

Who says so? Kathleen Deveny, "Firms See a Fat Opportunity in Catering to Americans' Quest for 'Easy' Lunches," *The Wall Street Journal*, November 3, 1992, p. B1.

Pizza State University

*a*ccording to a study done by Domino's Pizza, colleges with less than 10,000 students order 12 percent more pizza than larger institutions; freshman dorm students order 15 percent more pizza than upperclassmen; and female dorm students order the most pizza.

Who says so? The Wall Street Journal, "You Can Call This Study Cheesy, but It Gives a Slice of Campus Life," September 14, 1992, p. B1.

38

ASTOUNDING AVERAGES!

The Great Plains
(and Pepperonis, and Specials, and . . .)

*a*ccording to the National Association of Pizza Operators, Americans as a whole consume an average of 90 acres of pizza a day.

Who says so? Your Health and Safety, "Nutrition: Pizza Has Pizzazz," October–November 1991, pp. 18–19.

39

Cereal Killers

*a*mericans consumed an average of 11.90 pounds of breakfast cereal per capita in 1993, more than anyone else in the world. By contrast, South Koreans consumed only 0.07 pounds per capita.

Who says so? U.S. Department of Commerce, International Trade Administration, *U.S. Industrial Outlook '92.* Washington, D.C.: U.S. Department of Commerce, 1992, p. 32-17. *Who said so first?* *United Nations Yearbooks of Demographics, Industrial Statistics, and International Trade.*

40

ASTOUNDING AVERAGES!

That's the Way the Cookie Crumbles

he United States consumed an average of 12.29 pounds of cookies and crackers per capita in 1993. This amount was considerably less than that of the Netherlands, which led the world with 58.28 pounds per capita.

Who says so? U.S. Department of Commerce, International Trade Administration, *U.S. Industrial Outlook '92*, Washington, D.C.: U.S. Department of Commerce, 1992, p. 32-17. *Who said so first?* United Nations Yearbooks of Demographics, Industrial Statistics, and International Trade.

The Meat Department

*I*n 1992 the average per capita consumption of red meat and poultry in the United States was 178 pounds. For a family of four this would equate to a half a steer, a whole pig, and 100 chickens per year.

Who says so? Alan Thein Durning, "Fat of the Land: We Can't Keep Eating the Way We Do," *USA TODAY* (magazine), November 1992, p. 25.

ASTOUNDING AVERAGES!

On Top of Spaghetti

*a*mericans consume an average of 19 pounds of pasta per person per year. By some estimates this is twice as much as was consumed twenty years ago. By contrast, Italians eat an average of 60 pounds per person each year. According to a survey by *Consumer Reports*, readers of that magazine eat pasta an average of once a week, usually topped with a red sauce.

Who says so? *Consumer Reports*, "Spaghetti and Spaghetti Sauce," May 1992, p. 322.

Can Do

Each year Americans eat an average of 150 pounds of canned food, or 11 percent of food consumed. Because of its popularity, some 1,400 different canned food items are produced each year; 37 billion metal and glass containers are packed in more than 1 billion cases.

Who says so? Louis Rukeyser, *Louis Rukeyser's Business Almanac.* New York: Simon & Schuster, 1991, p. 441.

And It All Sticks to the Roof of Your Mouth

T he average American eats 3.36 pounds of peanut butter each year, of which 47 percent is creamy, 33 percent is crunchy, and 18 percent is natural.

Who says so? USA TODAY, "Peanut Butter Choice is Creamy," April 19, 1993, p. 1D. *Who said so first?* The Adult Peanut Butter Lovers Fan Club, Peanut Advisory Board.

Baby, What an Appetite

*I*n 1992 American babies ate an average of 53 to 54 dozen jars of baby food per baby, up from the 47 to 49 dozen jars consumed in the mid-1980s, but still not as much as the 66 dozen jars consumed in 1972.

Who says so? American Demographics, "Baby Food is Growing Up," May 1993, p. 20.

Fiends for Caffeine

*E*ach day Americans consume an average of 227 milligrams of caffeine, which is the equivalent of two to three cups of coffee. Tea and cola drinks are other sources of caffeine. There are an average of 40 milligrams in a six-ounce cup of tea and 45 milligrams in a twelve-ounce can of soda.

Who says so? Elisabeth Rosenthal, "Headache? You Skipped Your Coffee," *The New York Times*, October 15, 1992, p. A10.

CONSUMING PASSIONS

Be Like Mike

*a*ccording to the *New Age & Sports Beverages in the US: 1992* report from Beverage Marketing Corporation, per capita consumption of sports drink was 1.2 gallons in 1991. Gatorade accounted for more than 80 percent of this market.

Who says so? Eric Sfiligoj, "Is Gatorade a Sleeping Giant?" *Beverage World,* August 1992, p. 32.

Cheers!

Match the country with its annual per capita wine consumption.

1. Argentina
2. Italy
3. France
4. Australia
5. United States

A. 2.1 gallons
B. 5.1 gallons
C. 13.1 gallons
D. 17.4 gallons
E. 21.2 gallons

(Answers: 1C, 2E, 3D, 4B, 5A)

Who says so? Jobson's Wine Marketing Handbook, 1991. New York: Jobson Publishing, 1991, p. 131.

That's Oil, Folks!

*I*n developing countries, the average person uses 2 barrels of oil per year for fuel; the average Europeans and Japanese, 10 to 30 barrels per year; and the average Americans, 40 barrels per year.

Who says so? *Economist*, "Survey: Energy and the Environment: A Power for Good, a Power for Ill," August 31, 1991, p. 44.

A New Wrinkle on Energy Conservation

The average household clothes iron consumes 1 kilowatt of power for 2 hours of usage. Approximately 146 million irons in household use worldwide for 2 hours per week would equal ⅓ the total annual net electricity consumption of Switzerland, or ½ that of Greece.

Who says so? Vitali Matsarski, "How Crumpled Clothes Could Save the World," *New Scientist,* October 3, 1992, p. 46.

Water, Water Everywhere

The average American uses 25 to 30 gallons of water to shower, 2 gallons to brush their teeth, 10 to 15 gallons to shave, 20 gallons to wash dishes, 10 gallons to run the dishwasher, and 5 to 7 gallons to flush the toilet. The average individual uses 168 gallons of water daily and the average residence uses 107,000 gallons of water a year.

Who says so? The World Almanac and Book of Facts 1992. New York: World Almanac, 1991, p. 662. *Who said so first?* American Water Works Association.

Royal Flush

*a*n average of 4.8 billion gallons of water are used every day for flushing toilets in America.

Who says so? Garbage, January/February 1990, p. 12. *Who said so first?* U.S. Department of Housing and Urban Development; and *Garbage.*

Water Daily

*T*he average person needs 2 ½ quarts of water each day, 1 ½ quarts of which are obtained from liquids; the other quart comes from the water content of food, in particular fruits and vegetables. The percentage of water in selected types of food: bread, 35 percent; meat, 50 to 70 percent; pineapple, 87 percent; and a ripe tomato, 95 percent.

Who says so? Barbara Tufty, *1001 Questions About Earthquakes, Avalanches, Floods and Other Natural Disasters.* New York: Dover Publications, 1978, p. 245.

ASTOUNDING AVERAGES!

Just Add Water

*I*t takes an average of 32 gallons of water to produce one pound of steel; 280 gallons of water to produce one Sunday newspaper; 300 gallons of water to produce one pound of synthetic rubber; 1,000 gallons of water to produce one pound of aluminum; and 100,000 gallons of water to produce one automobile.

Who says so? William P. Cunningham and Barbara Woodhouse Saigo, *Environmental Science: A Global Concern.* Dubuque, IA: William C. Brown Publishers, 1999, p. 200.

55

CONSUMING PASSIONS

Just Add Water, Part II

*I*t takes an average of 40 gallons of water to produce one egg; 80 gallons of water to produce one ear of corn; 160 gallons of water to produce one loaf of bread; and 2,500 gallons of water to produce one pound of beef.

Who says so? William P. Cunningham and Barbara Woodhouse Saigo, *Environmental Science: A Global Concern.* Dubuque, IA: William C. Brown Publishers, 1990, p. 298.

ASTOUNDING AVERAGES!

Modern Times

Deep in the Red

*T*he national debt averaged only $16.60 per capita In 1900;
in early 1993 it was approximately $18,000 per capita.

Who says so? U.S. News & World Report, February 1, 1993, p. 13.

59

That's a Wrap!

The number of Christmas gifts wrapped each year by the average household is 30.

Who says so? *Garbage,* November/December 1990, p. 63. *Who said so first?* Hallmark Cards, A.C. Nielsen, and the National Christmas Association.

ASTOUNDING AVERAGES!

Season's Greetings

*I*n 1991 an average of 44 Christmas cards were sent per family, of which 41 percent were sent to non-relatives.

Who says so? U.S. News & World Report, "Database," December 21, 1992, p. 26.

A Letter Carrier's Nightmare

*a*ccording to the United States Postal Service, between Thanksgiving and Christmas Americans send out more than 2.3 billion holiday cards, or an average of 78 million cards per day.

Who says so? Catherine S. Manegold, "Rituals to Match the Ways People Live Now," *The New York Times*, December 24, 1992, p. C1.

ASTOUNDING AVERAGES!

Mail-Order Mania

*I*n 1981 an average of 59 mail-order catalogs were received per household; in 1991, the average number increased to 142. Approximately 55 percent of adults shop by mail or phone.

Who says so? U.S. News & World Report, "Database," December 21, 1992, p. 26.

63

Behind Closed Doors

ccording to the World Health Organization, there are more than 100 million acts of sexual intercourse each day.

Who says so? *The New York Times*, "U.N. Agency on Sex: Pitfalls and Promise," June 25, 1992, p. A12.

ASTOUNDING AVERAGES!

Why Must I Be a Teenager in Love

*I*n the United States each year an average of 1 out of 10 teenage women and 30,000 girls under the age of fifteen become pregnant. Each day an average of more than 3,000 girls become pregnant; 1,300 babies are born to teenage girls; 500 girls have induced abortions; and 13 sixteen-year-old girls have their second child.

Who says so? Hutzel Hospital, "Statistics," Detroit, MI.

It's Not the Size That Counts

*T*he average Japanese home is 650 square feet, or less than ½ the size of the average American home.

Who says so? Hal Foster, "Small-Scale Success Story," *Los Angeles Times,* July 15, 1991, p. D1.

Wet Paint

*a*n ordinary paint roller covers an average of 650 square feet with one gallon of paint, although manufacturers recommend 400 to 450 square feet for every gallon to obtain a "one coat" coverage.

Who says so? Consumer Reports, "New Paint Roller Lays it on Thick," October 1992, p. 626.

67

I'll Call You Back (Much) Later

*a*ccording to a report by the Geneva-based International Telecommunications Union, the average waiting period to start telephone service in Russia is 32 years. Conditions are far worse in Albania, where it would take 100 years to install telephones just for the people who have already requested service.

Who says so? Richard L. Hudson, "Thirty-two Years is the Average Wait for Phone in Russia," *The Wall Street Journal,* October 2, 1992, p. A5B.

68

ASTOUNDING AVERAGES!

It's a Crime

*I*n the United States a property crime occurs every 2 seconds; a larceny-theft every 4 seconds; a burglary every 10 seconds; a violent crime every 17 seconds; a motor vehicle theft every 19 seconds; an aggravated assault every 29 seconds; a robbery every 46 seconds; a forcible rape every 5 minutes; and a murder every 21 minutes.

Who says so? U.S. Department of Justice, Federal Bureau of Investigation, *Crime in the United States 1991.* Washington, D.C.. U.S. Department of Justice, 1992, p. 4.

Follow That Cart!

*I*n 1988 an average of 11 shopping carts were stolen per supermarket.

Who says so? The Food Marketing Institute, *The Food Market Industry Speaks.* Washington, D.C.: The Food Marketing Institute, 1989.

Attention Purse Snatchers:
You Already Know This

*a*ccording to a *Glamour* magazine survey on handbags, 50 percent of the women surveyed carried a shoulder bag which weighed an average of 3 to 5 pounds; 29 percent of the bags weighed under 3 pounds and 16 percent weighed over 5 pounds. The average handbag contained keys, wallet, checkbook, calendar, glasses, lipstick, pen, and gum or mints.

Who says so? *Glamour*, "How Heavy Is Your Bag and What's In It?" December 1992, p. 126.

Born to Be Wild

*a*ccording to the Motorcycle Industry Council, the average motorcyclist is a married male, age 32.5 years (up from the 1980s' average of 27.5 years), with some college education. His household income is $33,200, 12 percent higher than the national average.

Who says so? American Demographics, "Bikers Ride into Middle Age," December 1991, p. 15.

ASTOUNDING AVERAGES!

Born to Run (a Company)

*T*he average CEO is a white male protestant who has been with his company 21 years and, at age 56, has been a CEO for 8 years. He is a college graduate, but not usually a varsity type. Over ½ of all CEOs have received graduate degrees. In 1991 the average CEO made $878,000 in salary and bonuses. Their favorite hobbies are reading, collecting, and gardening.

Who says so? Laurel Touby and Fred F. Jespersen, "Portrait of the Boss," *Business Week*, October 12, 1992, pp. 108–109.

Franchise Player

*a*ccording to a survey by Francorp, a consulting firm in Olympia Fields, Illinois, the average franchisee is a 40-year-old white male with some college education and an annual income of $50,000–$70,000.

Who says so? Dan Frost and Susan Mitchell, "Small Stores with Big Names," *American Demographics,* November 1992, pp. 52–57.

ASTOUNDING AVERAGES!

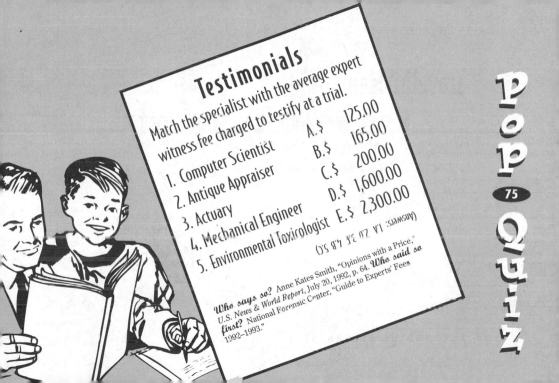

Testimonials

Match the specialist with the average expert witness fee charged to testify at a trial.

1. Computer Scientist
2. Antique Appraiser
3. Actuary
4. Mechanical Engineer
5. Environmental Toxicologist

A. $ 125.00
B. $ 165.00
C. $ 200.00
D. $ 1,600.00
E. $ 2,300.00

(Answers: 1.A 2.C 3.E 4.B 5.D)

Who says so? Anne Kates Smith, "Opinions with a Price," U.S. News & World Report, July 20, 1992, p. 64. **Who said so first?** National Forensic Center, "Guide to Experts' Fees 1992–1993."

Pop Quiz

75

Travelin' Man

*T*he average business traveler is a married 39-year-old male who does managerial or professional work and has a median household income of $40,000 a year.

Who says so? Deborah Schoeder and Judith Waldrop, "How Business Travelers Are Changing," *American Demographics*, November 1992, p. 10.

ASTOUNDING AVERAGES!

The Gambler

*a*ccording to a survey of the general public and casino gamblers, the average visitor to a casino—who is just as likely to be a man as a woman—is between the ages of 40 and 64, lives in a one- or two-member household, has a median household income of $35,000 a year (higher than the average American), and is slightly better educated than the average American.

Who says so? *The Wall Street Journal,* "Survey Reveals a Recent Jump in Casino Goers," January 8, 1993, p. B1.

77

Dad, Can I Borrow the Rolls?

*I*n 1983 the average net worth of America's wealthy elite (approximately 500,000 families) was $5.86 million per family, or 24.1 percent of total household wealth. In 1989 this average net worth increased to $10.3 million, or 29.1 percent of household wealth.

Who says so? Business Week, "The Superrich Have Been on a Roll," May 18, 1992, p. 22. *Who said so first?* Arthur B. Kennickell, Federal Reserve Board; and R. Louise Woodbaum, Internal Revenue Service.

ASTOUNDING AVERAGES!

Perks Plus

*I*n 1991 each member of the House of Representatives received an average allowance of $176,900 for office expenses and an average of $200,000 for mailing expenses. In addition, the average member's payroll was $479,900 and franked mail expenses were $100,000. Total spent on office expenses, payroll, and franked mail was an average of $734,700 per member.

Who says so? Anne Willette, "Lawmakers' Pay, Perks Add to Total," *USA TODAY*, September 28, 1992, pp. 1A–5A.

79

I'm Sorry, Senator Dole Has 83,999 Other Calls Waiting

The Capitol switchboard, which handles phone calls for 535 senators and representatives and 5 congressional delegates, receives 84,000 phone calls on an average day.

Who says so? Christopher Scanlan, "Americans Burn up Capital's Phone Lines," *Detroit Free Press,* January 28, 1993, p. 9A.

ASTOUNDING AVERAGES!

Mr. Popularity

*I*n 1993 the White House received a daily average of 75,850 pieces of correspondence for President Clinton, consisting of 250 telegrams, 600 computer messages, 25,000 letters, and 50,000 phone calls.

Who says so? USA TODAY, "Public Clamors for Clinton's Ear," April 2–4, 1993, p. 1A. *Who said so first?* The White House, Western Union

No Laughing Matter

*I*n the United States a child runs away from home every 26 seconds; a teenager has a baby every 67 seconds; a child is arrested for drug offenses every 7 minutes; a child drops out of school every 8 minutes; 100,000 children are homeless every day; and 135,000 children bring guns to school every school day.

Who says so? Marian Wright Edelman, "Kids First!" *Mother Jones,* May–June 1991, p. 31.

ASTOUNDING AVERAGES!

Time and Time Again

Don't Unpack Those Boxes Just Yet

The average American moves 11 times in a lifetime, or once every 6 years.

Who says so? Joseph F. Coates and others, "Future Work," *Futurist,* May–June 1991, p. 13.

TIME AND TIME AGAIN

It's Our Dream House! (or is it?)

*I*n 1992 the average buyer looked at 15.6 houses before making a purchase.

Who says so? Chicago Title and Trust Family of Title Insurers, Chicago, IL, *Who's Buying Homes in America,* 1993, p. 3.

ASTOUNDING AVERAGES!

Then Why Is the Mortgage for Thirty Years?

To earn enough money to purchase a house Americans had to work an average of 1,125.5 days in 1962; 1,330.7 days in 1972; 1,530.0 days in 1982; and 1,777.3 days in 1992.

Who says so? Consumer Reports, "Has Our Living Standard Stalled," June 1992, pp. 392–393

Ken Not Included

*T*o earn enough money to purchase a Barbie doll Americans had to work an average of 81.1 minutes in 1962; 60.8 minutes in 1972; 33.2 minutes in 1982; and 31.3 minutes in 1992.

Who says so? *Consumer Reports,* "Has Our Living Standard Stalled," June 1992, pp. 392–393.

Hours to Go Before I Sleep

o earn enough money to buy a new mattress Americans had to work an average of 71.6 hours in 1962; 59.5 hours in 1972; 44 hours in 1982; and 60.7 hours in 1992.

Who says so? *Consumer Reports,* "Has Our Living Standard Stalled," June 1992, pp. 392–393.

Work 10, Children 1.9

*a*ccording to Priority Management Systems, a management consulting firm in Bellevue, Washington, parents spend an average of 1.9 hours per day with their children, but 10 hours working and commuting.

Who says so? Joan E. Rigdon, "For Jacinta and Sam Mathis, Having It All Means Doing It All, with Barely Enough Time to Rest," *The Wall Street Journal*, June 21, 1993, p. R13.

Heigh Ho, Heigh Ho, It's Off to Work We Go!

On a national average, commuting to work in metropolitan areas takes 23.2 minutes each day. In New York, it's 30.6 minutes; Washington, D.C., 29.5; Chicago, 28.1; Los Angeles, 26.4; and Houston, 26.1.

Who says so? U.S. News & World Report, "Demographics," August 10, 1992, p. 15.

Please Hold

*a*ccording to a survey of 150 large companies done by OfficeTeam, a Menlo Park, California, personnel agency, the average executive spends 15 minutes every day waiting on hold.

Who says so? *Detroit Free Press*, "Bottom Line: Holding Patterns," July 7, 1993, p. 1C.

ASTOUNDING AVERAGES!

I Know It's Here Somewhere

*a*ccording to a survey of 200 executives by Accountemps, office workers waste an average of 50 minutes per day trying to locate mislabeled, misfiled, or mis-placed items.

Who says so? Working Women, January 1993, p. 34.

And Besides, the Dog Ate My Car Keys

*I*n a 100-person office, the most frequent reasons given for taking sick days and the average number taken per year are: cold, 21 days; fractures, 23 days; sprains, 30 days; and flu, 76 days.

Who says so? *Detroit Free Press,* "Hello Boss . . . ," March 14, 1993, p. 3A. *Who said so first?* National Center for Health Statistics.

21 Years and Out

*a*ccording to a survey of 300 midsized and large companies by William M. Mercer, a benefits consulting firm in New York, the average worker retires at age 62 or 63 after 21 years with his or her employer. But by the year 2000, the average worker will not be able to afford retirement at this age, largely due to increasing health care costs.

Who says so? The Wall Street Journal, "Health Care Costs Becloud Plans for Retirement," June 9, 1993, p. B1.

The Honeymoon Is Over

*I*n 1992 the average length of a honeymoon was 8 days, remaining unchanged from 1987.

Who says so? *American Demographics*, "Business Reports," August 1992, p. 17.

ASTOUNDING AVERAGES!

'Til Death Do Us Part?

*I*n 1988 the median duration of a marriage was 7.1 years.

Who says so? U.S. Bureau of the Census, *Statistical Abstract of the United States 1992*. 112th edition. Washington, D.C.: U.S. Department of Commerce, 1992, p. 92. **Who said so first?** U.S. National Center for Health Statistics, *Vital Statistics of the United States,* annual; *Monthly Vital Statistics Report;* and unpublished data.

Shop 'til You Drop

*a*dults who visit large, regional shopping malls do so an average of 3.9 times per month; those who visit small shopping centers do so an average of 7.1 times per month.

Who says so? Chip Walker, "Strip Malls: Plain but Powerful," *American Demographics*, October 1991, pp. 48–50.

Attention Shoppers:
Ignore the Tantrum in Aisle 5

*a*ccording to a study conducted through personal inter-
views of 1,440 households in Texas, children accom-
pany their parents to stores an average of 2.50
times per week; girls go with their parents 2.65
times a week and boys 2.34 times.

Who says so? James U. McNeal, "Growing Up In the Market,"
American Demographics, October 1992, p. 49.

TIME AND TIME AGAIN

But Longer If You're Hungry

*a*ccording to a study done by Market Growth Resources, a retail consulting firm, Americans are spending less time in the grocery store. Shopping which took an average of 28 minutes per trip in the early 1980s now takes an average of 22 minutes per trip.

Who says so? The Wall Street Journal, "Displays Pay Off for Grocery Marketers," October 15, 1992, p. B1.

ASTOUNDING AVERAGES!

Food for Thought

Match the grocery store department with the average amount of time spent by shoppers in that department.

1. Meat
2. Baby Food
3. Seafood
4. Produce
5. Bread
6. Liquor

A. 40.5 seconds
B. 42.3 seconds
C. 130.9 seconds
D. 136.6 seconds
E. 154.0 seconds
F. 181.0 seconds

(Answers: 1.E 2.C 3.A 4.F 5.B 6.D)

Who says so? Business Week, "Attention Shoppers," April 20, 1992, p. 44. *Who said so first?* Videocart.

I Thought This Was the Express Lane!

CNN and ActMedia's Checkout TV Channel showed news, features, and ads and was targeted to those waiting in checkout lines. Its 8-minute cycle coincided with the average time a grocery store shopper stands in line.

Who says so? American Demographics, "TV Takes on Tabloids at Checkout Line," April 1991, p. 9.

Beautiful, Man

*a*ccording to a survey done by *Gentlemen's Quarterly*, in 1990 men spent an average of 44 minutes per day arranging their hair and clothes, an increase of 30 minutes per day over 1988. Men under 25 years old spend the most time, an average of 53 minutes daily, working on their outward appearance.

Who says so? Diane Crispell, "The Brave New World of Men," *American Demographics*, January 1992, p. 42.

Bah Humbug!

*R*esults of a three-year study done by the Center for Lifestyle Management: only 9 minutes are spent by the average parent playing with his or her children Christmas morning. Respondents also revealed that an average of 10 hours are spent the last week of December arguing and bickering with family members about holiday-related activities.

Who says so? Detroit Free Press, "Christmas? Friends? Bah Humbug," December 20, 1992, p. 6A.

ASTOUNDING AVERAGES!

Vital Information

Happy Birthday, Dear Sammy (and Donna and Marc and Michele ...)

On an average day in the United States, some 3 million birthday gifts are purchased for the 673,693 people who are celebrating their birthday. An average of 1 in 68 individuals turns 18 every day; 1 in 61 turns 40.

Who says so? Les Krantz, *What the Odds Are: A to Z Odds on Everything You Hoped or Feared Could Happen.* New York: Prentice Hall General Reference, 1992, p. 41.

VITAL INFORMATION

Weight Watching

*T*he median birth weight of babies born in the United States to U.S. residents was 7 pounds, 7 ounces in 1989. This marked the tenth consecutive year that that figure has remain unchanged.

Who says so? U.S. Bureau of the Census, *Statistical Abstract of the United States 1992.* 112th edition. Washington, D.C.: U.S. Department of Commerce, 1992, p. 68. *Who said so first?* U.S. National Center for Health Statistics, *Vital Statistics of the United States,* annual; *Monthly Vital Statistics Report;* and unpublished data.

ASTOUNDING AVERAGES!

The Big Plunge

*I*n 1992 the average age of a first time home buyer was 31.0; the average age of a repeat buyer was 40.8.

Who says so? Chicago Title and Trust Family of Title Insurers, *Who's Buying Homes in America*. Chicago, IL, 1993, p. 3.

109

I Do

*I*n 1990 the median age at first marriage for men was 26.1 years, the same as it was in 1890. The median age at first marriage for women in 1990 was 23.9 years, up slightly from 22.0 years in 1890.

Who says so? The World Almanac Book of Facts 1992. New York: World Almanac, 1991, p. 943. *Who said so first?* U.S.Bureau of the Census.

I Don't

*J*n 1988 the median age at divorce was 35.1 years for men and 32.6 years for women.

Who says so? U.S. Bureau of the Census, *Statistical Abstract of the United States 1992.* 112th edition. Washington, D.C.: U.S. Department of Commerce, 1992, p. 92. *Who said so first?* U.S. National Center for Health Statistics, *Vital Statistics of the United States,* annual; *Monthly Vital Statistics Report;* and unpublished data.

The Young and the Restful

*P*rovo-Orem, Utah, with a median age of 21.5 years, has the youngest population in the United States; Sarasota, Florida, with an median age of 51.1 years, has the oldest.

Who says so? Les Krantz, *The Best and Worst of Everything*. New York: Prentice Hall General Reference, 1991, p. 66.

You're Not Getting Better, You're Getting Older

*I*n 1820 the median age of the United States' population was 16.7 years; in 1980 it was 30.0; and in 1990 it was 33.0. In the year 2010 the median age is expected to be 39.0 years, and by the year 2030, 41.0.

Who says so? *The Universal Almanac 1992.* Kansas City, MO: Andrews and McMeel, 1991, p. 208. *Who said so first?* U.S. Bureau of the Census.

113

Old World Charm

ecause of rising life expectancy and declining birth rates, the average age of the population in many countries is increasing. By the year 2040, the segment of population over 65 years of age in the United States is expected to increase around 75 percent; in Japan a 149 percent increase is expected.

Who says so? William E. Schmidt, "Anxiety in a Graying Europe," *International Herald Tribune,* July 14, 1993, p. 5.

ASTOUNDING AVERAGES!

Getting Mighty Crowded

Match the city with its population per square mile.

1. Cairo
2. Hong Kong
3. New York City
4. Bangkok
5. Bombay

A. 11,480
B. 58,379
C. 97,106
D. 127,461
E. 247,501

(Answers: 1C, 2E, 3A, 4B, 5D)

Who says so? U.S. Bureau of the Census, *Statistical Abstract of the United States 1992.* 112th edition. Washington, D.C.: U.S. Department of Commerce, 1992, pp. 825–826. *Who said so first?* U.S. Bureau of the Census, *World Population Profile: 1991.*

Pretty Soon
We Won't Need Any Houses

The size of the average American household was 5.8 persons in 1790; 4.8 persons in 1900; 3.7 persons in 1940; 3.3 persons in 1950; and 2.6 persons in 1990.

Who says so? Diane Crispell, "Myths of the 1950s," *American Demographics*, August 1992, pp. 38–43.

ASTOUNDING AVERAGES!

And Baby Makes 3.18

*I*n 1991 the average U.S. household consisted of 2.63 persons and the average U.S. family consisted of 3.18 persons.

Who says so? U.S. Bureau of the Census, *Statistical Abstract of the United States 1992.* 112th edition. Washington, D.C.: U.S. Department of Commerce, 1992, p. 46. ***Who said so first?*** U.S. Bureau of the Census, *Current Population Reports,* series P-20, No. 458.

117

Honey, I Shrunk the Kids

*I*n 1991 the average number of children per American family was .96, dropping from 2 children for a typical family in 1977. Among America's 24 million two-parent families with children, there are an average of 1.88 "own" children (householder's biological sons and daughters, stepchildren, and adopted children) living in each household.

Who says so? *American Demographics*, "Fractional Families," December 1992, p. 6.

And Some Guys *Still* Can't Get a Date!

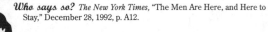

ccording to the Census Bureau, the male-female ratio in the United States is an average of 95 men to 100 women.

Who says so? The New York Times, "The Men Are Here, and Here to Stay," December 28, 1992, p. A12.

119

VITAL INFORMATION

Words to Live By

*T*he following are the most common words used by the average person who speaks English: the, of, and, to, a, in, that, is, I, it, for, as.

Who says so? Gyles Brandreth, *Your Vital Statistics.* New York: Citadel, 1986, p. 50.

ASTOUNDING AVERAGES!

Well, Maybe When They're 50 ...

*T*he average 11-year-old reads only 11 pages of text per day.

Who says so? *Advertising Marketing Bulletin,* "New Computerization," Spring 1993, p. 10.

Now, Use Each One in a Sentence

The vocabulary of the average high school graduate is said to consist of 50,000 words.

Who says so? *Consumer Reports,* "Spell Checkers: The Dictionaries of Tomorrow?" October 1991, p. 672.

ASTOUNDING AVERAGES!

Testing, 1, 2, 3

*I*n 1992 the average Scholastic Aptitude Test (SAT) score was 899 out of a possible 1,600 points. This is down from the 1972 average score of 937. The first year that the College Board compiled average scores was 1972, and the scores haven't been as high since.

Who says so? USA TODAY, "Putting Them to the Test," August 27, 1992, p. D1. *Who said so first?* The College Board.

123

Check It Out

*a*ccording to a study of 8,978 public libraries across the United States, there were an average of 3.13 visits per capita to public libraries in 1990, or 507 million visits nationwide. Each visit resulted in the checking out of almost 2 items.

Who says so? Library Journal, "Government Releases Telling Data on U.S. Public Libraries," October 15, 1992, p. 16. *Who said so first?* National Center for Education Statistics, "Public Libraries in the U.S.: 1990"

124

ASTOUNDING AVERAGES!

Alien Nation

*T*he Census Bureau estimates that legal and illegal immigration combined will increase the United States' population an average of 800,000 every year for the next six decades. This group of immigrants and their offspring will account for 21 percent of the population by the year 2050.

Who says so? Los Angeles Public Library/State of California, *Scan/Info,* December 1992, pp. 16–18. *Who said so first?* *The New York Times,* December 4, 1992, pp. A1, A4.

125

Habla Español?

*I*n the Miami metropolitan area, 41.6 percent of the households speak a foreign language, the highest such percentage in the United States. Rounding out the top five are Los Angeles, 38.3 percent; New York, 28.4 percent; San Francisco, 26.8 percent; and San Diego, 25.0 percent. The national metro average is 15.8 percent.

Who says so? U.S. News & World Report, "Demographics," August 10, 1992, p. 15. *Who said so first?* U.S. Census, 1990.

Taking Care of
Business

... and After That,
I Was President of the United States

hirty percent of all resumes are falsified, according to Team Building Systems, an employee screener in Houston, Texas. Businesses spend $100 for an average background check of a new hire.

Who says so? *Inc.*, "Screening New Hires," August 1992, p. 82.

TAKING CARE OF BUSINESS

Don't Call Us, We'll Call You

*a*n average of 1,000 unsolicited resumes arrive daily at the *Fortune* 500 companies. After a quick look, 4 out of 5 are discarded.

Who says so? U.S. News & World Report, "Writing a Computer-Friendly Resume," October 26, 1992, p. 90.

We're Sorry, All Circuits Are Busy

here are about 9.5 billion minutes of telephone calling on an average business day.

Who says so? U.S. Department of Commerce, Industrial Trade Administration, *U.S. Industrial Outlook '92*. Washington, D.C.: U.S. Department of Commerce, 1992, pp. 28–29.

131

TAKING CARE OF BUSINESS

I Said I'm Not Interested! (Click)

 n an average day, over 300,000 telemarketers call more than 18 million Americans.

Who says so? Larry Pressler, "Statements on Introduced Bills and Joint Resolutions," *Congressional Record,* July 10, 1991, p. S9498.

ASTOUNDING AVERAGES!

Just the Fax, Ma'am

*T*he average *Fortune* 500 company received 428 pages per day by fax in 1993, compared with 300 a day in 1992; it sent an average of 260 pages (49 documents) per day in 1993, an increase of 41 percent over 1992.

Who says so? *USA TODAY,* "Fax Users Paper Offices," May 18, 1993, p. 2B.

133

Please Print Legibly

*a*ccording to the Writing Instrument Manufacturers Association, businesses lose $200 million per year because of poor handwriting. At the Los Angeles Post Office alone, 12 handwriting experts try to decipher 500 pieces of mail per day. Eastman Kodak is unable to return an average of 400,000 rolls of film per year because of illegible addresses.

Who says so? John Maines, "Getting It Write the First Time," *American Demographics*, December 1992, p. 18.

ASTOUNDING AVERAGES!

Hey, This Thing Plays Solitaire!

*a*ccording to a study by SBT Accounting Systems, workers spend an average of 5.1 hours per week "fiddling" with their PCs. Research done by 3M shows that 30 percent of PC users lose data and spend an average of one week reconstructing it. This loss of productivity amounts to 24 million days per year nationally, at a cost of $4 billion.

Who says so? William M. Bulkeley, "Data Trap: How Using Your PC Can Be a Waste of Time, Money," *The Wall Street Journal*, January 4, 1993, p. B5.

Occupant (or Current Resident)

In the United States, 63.5 billion pieces of bulk mail (commercial and nonprofit solicitations) and 13.6 billion catalogs are sent out each year. This amounts to an annual average of more than 300 pieces of "junk" mail for each person living in the U.S.

Who says so? Elizabeth Brown, "Ideas: How to Pare Your Share of 77 Billion Pieces of Junk Mail," *The Christian Science Monitor,* January 15, 1991, p. 14.

ASTOUNDING AVERAGES!

Overexposure

**a**mericans were exposed to an average of 1,500 advertising messages each day in the 1960s. In the 1990s they are exposed to 3,000 per day.

Who says so? *Consumer Reports,* "Advertising Everywhere!" December 1992, p. 752.

Now Playing at a Theater Near You

*I*n 1992 a film running for several months had an average advertising and promotion cost of $12 million.

Who says so? Geraldine Fabrikant, "Warner's Sequel Weapon Cuts Down Promotion Costs," *The New York Times*, May 12, 1992, p. D23.

ASTOUNDING AVERAGES!

Let's See That Again

a movie that earns more than $100 million at the box office will do well in video stores, selling an average of 8 copies at each video specialty outlet.

Who says so? *American Demographics,* "VCR Wars," August 1992, p. 6.

I Need a New Drug

a new drug takes an average of 12 years and $230 million to develop. For every successful drug that reaches the pharmacy, 10,000 compounds are duds.

Who says so? Newsweek, "Penicillin From a Screen," September 14, 1992, p. 58.

140

Hallmark Holidays

Match the holiday with the average number of greeting cards sold for that holiday.

1. Valentine's Day   A. 40 million
2. Easter   B. 101 million
3. Christmas   C. 150 million
4. Thanksgiving   D. 165 million
5. Mother's Day   E. 1.0 billion
6. Father's Day   F. 2.3 billion

(Answers: 1.E 2.D 3.F 4.A 5.C 6.B)

Who says so? Business Week, "Or Better Yet, Give Him a Tie," June 22, 1992, p. 44. *Who said so first?* Greeting Card Association.

Color My World

*E*ach year more than 2 billion Crayola crayons are produced, or an average of 5 million daily, enough to circle the globe 4 ½ times.

Who says so? Binney & Smith, a division of Hallmark Cards.

Calling All Teenagers

*a*t 4.2 million square feet, the Mall of America near Minneapolis was the largest enclosed shopping center in the United States in 1992. For it to be successful, every potential shopper must visit an average of 4 times per year and stay an average of 4 hours per visit.

Who says so? Judith Waldrop, "The Biggest Mall of All," *American Demographics*, August 1992, p. 4.

You Can Make Book on It

*O*n an average workday at the Library of Congress, 730 books are cataloged; 1,900 inquiries are answered in its Copyright Office; 2,000 responses to research assignments for Congress are handled in its Congressional Research Service Office; and $1.2 million dollars of taxpayers money is spent.

Who says so? Linton Weeks, "*The Washington Post* Looks at the Library," *Library of Congress Information Bulletin,* September 9, 1991, p. 327.

ASTOUNDING AVERAGES!

Look for the Union Label

*a*ccording to 1990 data compiled by the Bureau of Labor Statistics, membership in trade unions was worth $119 more in the average worker's weekly paycheck. Full-time employees with a unionized wage or salary had median weekly earnings of $509, compared to $390 for nonunion workers.

Who says so? John R. Oravec, "Economy: Union Pay Well Ahead of Nonunion," *AFL-CIO News*, February 18, 1991, p. 10.

145

But Our Parts Are Cheaper

*T*he average labor cost per vehicle produced by General Motors is $2,388; by Chrysler, $1,872; and by Ford, $1,629. The average labor cost per vehicle produced by Honda is $920.

Who says so? U.S. News & World Report, "Hitting the Brakes Hard," November 9, 1992, pp. 78–79. *Who said so first?* Economic Strategy Institute, Harbour & Associates, and the WEFA Group.

Have You Driven a
Health Care Plan Lately?

*I*n 1993 the cost of health care for employees at the Ford Motor Company averaged $800 per automobile. This is double the per-vehicle cost of steel.

Who says so? The Detroit News and Free Press, "Costing the Big Three Billions," March 21, 1993, p. 1A.

Down on the Farm

*T*he average farm in the United States is 467 acres, compared with 3 acres for the average Japanese farm.

Who says so? William Cook, David Lawday, and Jim Impoco, "Sowing Seeds for a Global Trade War," *U.S. News & World Report*, November 23, 1992, p. 72.

ASTOUNDING AVERAGES!

Put This in Your Pipe and Smoke It

Tobacco is the United States' most profitable crop, partly due to a government price-support program. According to the Department of Agriculture, in 1991 farmers earned a gross average income of $3,862 for every acre of tobacco planted, but only $691 for peanuts, $380 for cotton, $262 for feed corn, and $101 for wheat.

Who says so? Ronald Smothers, "Tobacco Country Quaking Over Cigarette Tax Increase," *The New York Times,* March 22, 1993, p. A1.

TAKING CARE OF BUSINESS

The Dope on Opium Farming

*I*n a country where the average annual per capita income is only $70, farmers in Afghanistan can earn an average of $200 per year from an opium harvest. This is ten times what he could earn from wheat. An Afghan farmer receives an average of $22 per pound for raw opium, but by the time it reaches the United States it sells for $40,000 to $100,000 per pound.

Who says so? Klaus Reisinger, *U.S. News & World Report*, "After War, a Deadly Harvest," August 17, 1992.

ASTOUNDING AVERAGES!

A Bargain at Twice the Price

*a*ccording to a study by Cornell University, household labor done by married women has an average value of $5.50 per hour after taxes ($7.64 per hour before taxes), or $10,000 annually.

Who says so? *The Christian Science Monitor,* "Housework of Married Women Worth $5.50 an Hour After Taxes," January 13, 1993, p. 7.

Situation Wanted

*T*here are an average of 1,500 corporate jobs lost every day in the United States.

Who says so? Marguerite Smith and Debra Englander, "The Best Places to Live in America," *Money*, September 1992, p. 117.

152

Body Language

Don't Cell Yourself Short

The average person is made up of 50 trillion cells. Contrast this with the average elephant, which has 6.5 quadrillion cells, or the average shrew, which has 7 billion cells.

Who says so? Isaac Asimov, *The Human Body.* New revised edition. New York: A Mentor Book, 1992, p. 312.

. . . The Hip Bone's Connected to the Thigh Bone . . .

*a*lthough babies are born with about 350 bones, an average adult has 206 bones because many of them fuse together between birth and maturity. Half of these are found in the hands and feet. The longest bone is the femur and it makes up 27 percent of the height of the average person.

Who says so? Gyles Brandreth, *Your Vital Statistics.* New York: Citadel, 1986, p. 14.

The Eyes Have It

The muscles operating the lenses in the eye move an average of 100,000 times per day focusing on various objects. These muscles work harder than any others in the body; to give leg muscles the equivalent amount of work would require walking 50 miles per day.

Who says so? Gyles Brandreth, *Your Vital Statistics.* New York: Citadel, 1986, p. 14.

In the Blink of an Eye

The rate of blinking varies, but on average the eye blinks once every five seconds, 17,000 times each day, and 6,205,000 times a year.

Who says so? Gyles Brandreth, *Your Vital Statistics.* New York: Citadel, 1986, p. 26.

ASTOUNDING AVERAGES!

Hair Today, Gone Tomorrow

Beneath the human scalp, a follicle grows 1 hair for 2 to 6 years at an average rate of ½ inch per month. At that point, the follicle shrinks toward the skin's surface. After resting for about 6 months, the follicle then sheds the hair. The average adult sheds 100 hairs a day.

Who says so? Consumer Reports, "What Is Hair," July 1992, p. 398.

A Hair-Raising Statistic

There are about 100,000 hairs on the average adult's scalp. Most redheads have about 90,000 hairs, blonds have about 140,000, and brunets fall in between these two figures.

Who says so? Margo, *Growing New Hair.* Brookline, MA: Autumn Press, 1980, p.40.

ASTOUNDING AVERAGES!

The Five O'Clock Shadow Explained

On the face of the average adult male, 25,000 hairs grow up to ½ millimeter every 24 hours.

Who says so? *How in the World?* Pleasantville, NY: The Reader's Digest Association, 1990, p. 29.

This Bud's for You

The average adult mouth has 10,000 taste buds in the throat, on the tongue, and on the roof of the mouth. The average child has slightly more.

Who says so? Neil McAleer, *The Body Almanac.* New York: Doubleday, 1985, pp. 62–63.

Spit in the Ocean

In the average lifetime, about 10,000 gallons, or more than 1.2 million fluid ounces, of saliva are produced by the average person—enough to fill a good-sized swimming pool. In the average person there are 45 ounces of saliva flowing in a 24-hour period.

Who says so? Neil McAleer, *The Body Almanac.* New York: Doubleday, 1985, p. 63.

163

Gimme Some Skin

he average man has about 20 square feet of skin, the average woman about 17 square feet. The outer skin is replaced about once a month.

Who says so? Gyles Brandreth, *Your Vital Statistics*. New York: Citadel, 1986, p. 16.

The Skinny on Skin

The skin of the average adult weighs about 5.9 pounds and, medically speaking, is considered to be the heaviest organ.

Who says so? Guinness Book of Records 1992. New York: Bantam Books, 1992, p. 183.

I've Got You Under My Skin

On average, 12 feet of nerves, 3 feet of blood vessels, 10 hair follicles, 100 sweat glands, and hundreds of nerve endings are in each square centimeter of skin. During the average lifetime 40 pounds of dead skin are shed.

Who says so? Gyles Brandreth, *Your Vital Statistics*. New York: Citadel, 1986, p. 16.

A Cell-ebration of Life

Match the human cell with its average lifespan.

1. Red blood cells
2. Bone cells
3. Spermatozoa
4. Brain cells
5. Skin cells

A. 2-3 days
B. 19-34 days
C. 120 days
D. 25-30 years
E. A lifetime

(Answers: 1.C 2.D 3.A 4.E 5.B)

Who says so? Diagram Group, *Comparisons*. New York: St. Martin's Press, 1980, p. 175; *Guinness Book of Records 1992*. New York: Bantam Books, 1992, p. 180.

But It Only Takes One

*a*ccording to a study in the *British Medical Journal,* average sperm concentrations had fallen to 66 million spermatozoa per milliliter in 1990. On average, men have about half the sperm counts of their grand-fathers' generation.

Who says so? *The Wall Street Journal,* "Human Sperm Count Study Shows World-Wide Decline," September 11, 1992, p. B13.

Eat Your Heart Out, Mr. Scarecrow

*T*he average human brain weighs 3 pounds and contains 10,000 million nerve cells, which have 25,000 potential interconnections with other cells.

Who says so? *The Guinness Book of Answers.* 8th edition. Enfield, England: Guinness Publishing, 1991, p. 148.

Nice Beat, but You Can't Dance to It

*I*n an average lifetime the human heart beats 3,000 million times, pumping 48 million gallons of blood around the body. In a regular rhythm it beats an average of 72 times per minute.

Who says so? Gyles Brandreth, *Your Vital Statistics.* New York: Citadel, 1986, p. 293.

170

Turn That Frown Upside Down

An average of 43 muscles are needed for a frown, but only 17 for a smile. In the average human body there are about 656 muscles.

Who says so? Gyles Brandreth, *Your Vital Statistics*. New York: Citadel, 1986, p. 26.

BODY LANGUAGE

Official Size and Weight

The average female is 5 feet 3¾ inches tall and weighs 135 pounds. The average male is 5 feet 9 inches tall and weighs 162 pounds. Between 1960 and 1990 the average American man grew 2 inches taller and 27 pounds heavier, while the average American woman grew 2 inches taller and 1 pound heavier.

Who says so? Isaac Asimov, *The Human Body.* New revised edition. New York: A Mentor Book, 1992, p. 314; Diagram Group, *Comparisons.* New York: St. Martin's Press, 1980, pp. 72–73.

Here She Comes, Miss America

etween 1954 and 1980, the weight of the average Miss America dropped from 132 pounds to 117 pounds.

Who says so? Kerry O' Neil, " 'The Famine Within' Probes Women's Pursuit of Thinness," *The Christian Science Monitor,* August 31, 1992, p. 11.

173

Not Tonight, I Have a Headache

*a*ccording to Dr. Seymour Solomon, director of the headache unit at Montefiore Medical Center in New York City, migraine patients have an average of 2 to 3 headaches per month, but many have them several times per week. An estimated 24 million Americans afflicted with migraines lose more than 157 million work days per year.

Who says so? Patricia Braus, "Migraine Misery: Whose Head Hurts Most," *American Demographics,* September 1992, p. 25.

ASTOUNDING AVERAGES!

Make Yourself Useful

The average human body contains enough iron to make a three-inch nail; sulphur to kill all fleas on the average dog; carbon to equal 900 pencils; potassium to fire a toy cannon; fat to make 7 bars of soap; phosphorus to make 2,200 match heads; and water to fill a ten-gallon tank.

Who says so? Gyles Brandreth, *Your Vital Statistics.* New York: Citadel, 1986, p. 14.

The Executive Workout

*W*orking quietly in his office, an executive uses an average of 105 calories per hour; a secretary burns up fewer, an average of 88 per hour.

Who says so? Gyles Brandreth, *Your Vital Statistics*. New York: Citadel, 1986, p. 113.

176

Kiss It Good-bye

*a*ccording to Italian nutritionists in an article in the *Environmental Nutrition* newsletter, the energy used in a kiss averages between 6 and 12 calories. At an average of 9 calories per kiss, 3 kisses per day for 365 days would burn up 9,855 calories.

Who says so? Alan Bluman, *Elementary Statistics.* Dubuque, IA: William C. Brown Publishers, 1992, p. 81.

BODY LANGUAGE

Laughter *Is* the Best Medicine

*a*ccording to Joan Coggin, M.D., a cardiologist at Loma Linda University School of Medicine in Loma Linda, California, laughter has proven health benefits—it leads to stress reduction and relaxation. Adults laugh an average of only 15 times per day, while children laugh an average of 400 times daily.

Who says so? *Glamour,* "Laugh—It's Healthy," December 1992, p. 33.

ASTOUNDING AVERAGES!

On the Road Again

Are We There Yet?

*T*he average American takes approximately 50,000 automobile trips in his or her lifetime.

Who says so? Les Krantz, *What the Odds Are: A to Z Odds on Everything You Hoped or Feared Could Happen.* New York: Prentice Hall General Reference, 1992, p. 27.

181

Morning Drive

*I*n 1985 the average one-way distance traveled to work was 10.73 miles and the average one-way commute time was 19.66 minutes.

Who says so? Motor Vehicle Manufacturing Association, *MVMA Motor Vehicle Facts and Figures '91*. Detroit, MI: Motor Vehicle Manufacturing Association, 1991, p. 52. *Who said so first?* U.S. Department of Energy, *Transportation Energy Data Book: Edition 11, 1991*.

ASTOUNDING AVERAGES!

Life in the Slow Lane

*I*n 1972 automobiles traveled along Los Angeles freeways at an average speed of 60 miles per hour; in 1982 they crawled along at an average of 17 miles per hour.

Who says so? Samia El-Badry and Peter K. Nance, "Driving Into the 21st Century," *American Demographics,* September 1992, p. 46.

ON THE ROAD AGAIN

Drive Faster!
That Snail Just Passed Us!

*I*n 1900 the average speed of a horse-drawn carriage was 8 miles per hour; in 1988 automobiles could travel no faster crossing central London. Other cities in the same predicament and their traffic's average speed: New York City, 9.9 miles per hour; Paris, 10.5 miles per hour; and Stockholm, 11.2 miles per hour.

Who says so? *How in the World?* Pleasantville, NY: The Reader's Digest Association, 1990, p. 50.

184

And You Thought L.A. Was Bad

*a*ccording to a report by the International Road Federation in Geneva, the city with the highest density of vehicles on its roads is Hong Kong, with 261 vehicles per 0.6 miles.

Who says so? *International Herald Tribune,* "Travel Update," November 19, 1992, p. 2.

Highways from Hell

The busiest highway in America is New York City's George Washington Bridge/I-95, with an average two-way traffic count of 249,300 vehicles per day. Chicago's Dan Ryan and Kennedy Express-ways (I-90/I-94) are a close second with an average of 248,000 vehicles per day.

Who says so? Les Krantz, *The Best and Worst of Everything.* New York: Prentice Hall General Reference, 1991, p. 273. *Who said so first?* Federal Highway Administration.

ASTOUNDING AVERAGES!

No Wonder There's So Much Traffic

*a*n average of 7,700 additional cars and trucks are crowding the roads of the United States every day.

Who says so? Dick Swett, member of the Public Works and Transportation Committee, "Can We Prevent a Grim Future for American Transportation?" *USA TODAY* magazine, March 1992, p. 14.

Hey, Can I Borrow Your Car?

*I*n China there is 1 car for every 822 people; in India, 1 car for every 408 people; and in Yemen, 1 car for every 314 people. In the United States there is 1 car for every 1.7 people.

Who says so? Motor Vehicle Manufacturing Association, *MVMA Motor Vehicle Facts and Figures '91*. Detroit, MI: Motor Vehicle Manufacturing Association, 1991, pp. 36–37.

Mass (Transit) Hysteria

Match the public transit system with its average number of passengers per year.

1. Paris
2. New York City
3. Mexico City
4. Tokyo
5. Moscow
6. Bombay

A. 2.426 billion
B. 1.694 billion
C. 1.407 billion
D. 1.156 billion
E. 1.117 billion
F. 1.006 billion

(Answers: 1.D, 2.F, 3.E, 4.B, 5.A, 6.C)

Who says so? Les Krantz, *The Best and Worst of Everything.* New York: Prentice Hall General Reference, 1991, p. 107. *Who said so first?* International Union of Public Transport, Brussels.

They Don't Drive 'em, They Just Export 'em

*a*utomobiles are used less in Japan than in the United States: an average of 4,400 miles per year in Japan vs. 9,500 miles per year in the U.S. The cars in Japan are also younger: an average of 4.8 years old in Japan vs. 7.6 years old in the U.S.

Who says so? Economist, "Business: Parts Makers Go Spare," July 6, 1991, p. 68.

Watch Out for the Other Guy

*a*n average of 5,223 people die as a result of automobile accidents each year in California, more than in any other state. Texas ranks second in this category with an average of 3,568 deaths per year.

Who says so? Les Krantz, *What the Odds Are: A to Z Odds on Everything You Hoped or Feared Could Happen.* New York: Prentice Hall General Reference, 1992, p. 27.

ON THE ROAD AGAIN

Would You Like the Insurance with That?

The average daily car-rental rate was $56 in 1992, down from $65 in 1989.

Who says so? USA TODAY, "Lower Cost is No Boon," September 8, 1992, p. 8E. *Who said so first?* Smith Travel Research.

Driven to Death

*a*n average rental car is driven 2,000 to 3,000 miles each month.

Who says so? James S. Hirsch, "Rental Car Firms Jack Up Their Prices," *The Wall Street Journal,* November 4, 1992, p. B1.

Now *That's* Full Service

 n a 500-mile race on the NASCAR circuit, the average stock car makes 3 pit stops to make adjustments to handling, to refuel, and to change tires. The average pit stop in the 1950s took 4 minutes; in the 1960s, about 1 minute; in 1990, less than 22 seconds.

Who says so? Joseph Blackburn, "Gentlemen, Start Your Production Lines," *The Christian Science Monitor,* August 30, 1991, p. 19.

Think of the Lost Luggage

One of the world's busiest airports, Chicago's O'Hare handles 55 million people every year, or an average of 6,700 passengers per hour. An average of 2,200 planes, from 50 commercial airlines, pass through O'Hare each day.

Who says so? How in the World? Pleasantville, NY: The Reader's Digest Association, 1990, p.32.

Fly Me to the Moon

he average cost of unrestricted coach airfare was 23 cents per mile in 1992, down from 26 cents in 1987 and 34 cents in 1991.

Who says so? USA TODAY, "Lower Cost is No Boon," September 8, 1992, p. 8E. *Who said so first?* Smith Travel Research.

Frequent Flier Miles

The average number of miles per airplane trip was 862 in 1991, down from 1,200 in 1987, 1,100 in 1988, and 1,000 in 1990.

Who says so? USA TODAY, "Lower Cost is No Boon," September 8, 1992, p. 8E. *Who said so first?* Smith Travel Research.

197

Have Money, Will Travel

The average daily cost of meals and lodging in New York City is $297, the highest of any U.S. city. Other cities in the top 10 are: Washington, D.C. ($248), Honolulu ($238), Chicago ($215), Boston ($197), San Francisco ($194), Atlantic City ($191), Los Angeles ($190), Anchorage ($167), and Philadelphia ($161).

Who says so? *Parade* magazine, "Big Apple Takes Big Bite Out of Travel Budget," January 24, 1993, p. 9. *Who said so first?* Runzheimer International, a Wisconsin-based consulting firm.

Room Service Not Included

*T*he average hotel room rate was $69 per night in 1992, up from $54 in 1987.

Who says so? *USA TODAY,* "Lower Cost is No Boon," September 8, 1992, p. 8E. ***Who said so first?*** Smith Travel Research.

199

Please Have Your Passports (and Credit Cards) Ready

On average, 41 foreign visitors arrive in the United States every minute. Foreign visitors to the U.S. spend an average of $22,000 each minute, during which they generate an average of $291 in local taxes, $851 in state taxes, and $1,392 in federal taxes.

Who says so? Louis Rukeyser, *Louis Rukeyser's Business Almanac.* New York: Simon & Schuster, 1991, p. 552.

ASTOUNDING AVERAGES!

That's
Entertainment

And the Answer Is...

The average winner on *Jeopardy!* earns $11,500 per show. Every year, out of the tens of thousands who inquire about becoming a contestant, an average of 20,000 people take the qualifying exam, from which about 450 actually make it onto the show.

Who says so? Anthony Cook, "Going for the Grand Prize," *Money*, September 1992, p. 131.

The Family That Watches Together...

*I*n 1989 American households spent an average of 7 hours and 1 minute per day watching television. In 1971 the average was 6 hours and 2 minutes per day.

Who says so? The Universal Almanac 1992. Kansas City, MO: Andrews and McMeel, 1991, p. 251. *Who said so first?* Television Bureau of Advertising, 1991.

ASTOUNDING AVERAGES!

Must Be Those Soap Operas

omen 55 and older watch the most television per week, an average of 38 hours and 53 minutes; teenage girls watch the least television per week, an average of 21 hours and 18 minutes.

Who says so? Les Krantz, *The Best and Worst of Everything.* New York: Prentice Hall General Reference, 1991, p. 45. *Who said so first?* Nielsen Media Research.

Hey, Kids!
Coming Up Next, More Violence!

*a*ccording to a report to the National Cable Television Association, there are an average of 5.2 violent acts per program and 17.3 per hour in children's programming created for cable; on broadcast television there are an average of 7.8 violent acts per program and 32 per hour.

Who says so? Diane Duston, "Kids Shows Most Violent, Research Finds," *Detroit Free Press,* January 28, 1993, p. 3A.

ASTOUNDING AVERAGES!

Killing Time Can Be Murder

*T*he average child in the United States sees 26,000 murders on television by his or her eighteenth birthday.

Who says so? *Guinness Book of Records 1992.* New York: Bantam Books, 1992, p. 510.

207

THAT'S ENTERTAINMENT

Kids See the Darndest Things

The average American teenager has viewed 22,000 hours of television (which includes 15.8 examples of sexual language or imagery per hour of prime time TV) by his or her high school graduation.

Who says so? U.S. News & World Report, "Long Island Lolita," January 4, 1993, p. 104.

ASTOUNDING AVERAGES!

Ad Nauseam

*A*merican teenagers are exposed to an average of 3 to 4 hours of television advertising per week. This amounts to around 100,000 ads between birth and high school graduation.

Who says so? Alan During, "Leftfield: Enough is Enough," *Dollars & Sense*, June 1991, pp. 15–18.

A Strong Case for Public Broadcasting

*T*he average child in the United States views 350 commericals on television each week.

Who says so? Marc Silver, *U.S. News & World Report,* "Troubling TV Ads," February 1, 1993, pp. 65–67.

ASTOUNDING AVERAGES!

We'll Be Back,
Right After These Messages

I n an average hour of prime time television there are 9 minutes of commercials, or about 15 percent of each hour.

Who says so? Kevin Goldman, "Barrage of Ads in Super Bowl Blurs Messages," *The Wall Street Journal,* February 3, 1993.

THAT'S ENTERTAINMENT

Stop, Look, Listen

*P*er capita, Americans spend an average of 3 hours and 48 minutes a day watching television; 3 hours and 21 minutes a day listening to the radio; 34 minutes a day reading newspapers; and 20 minutes a day reading magazines.

Who says so? *Fortune,* "Top Cities for Watching, Listening, and Reading," May 18, 1992, p. 18.
Who said so first? Young & Rubicam.

ASTOUNDING AVERAGES!

Radio Daze

*I*n an average hour, radio listeners hear 12.7 commercials—totaling 10 minutes and 6 seconds—on AM radio and 11.9 commercials—lasting 9 minutes and 17 seconds—on FM radio.

Who says so? Louis Rukeyser, *Louis Rukeyser's Business Almanac.* New York: Simon & Schuster, 1991, p. 509.

213

THAT'S ENTERTAINMENT

No Vacancy

*I*n any given week an average of 2.3 million Americans are on paid vacation. During June, 3.3 million Americans are on paid vacation per week; July, 5.9 million; and August, 5.7 million.

Who says so? USA TODAY, "Vacation Time is Here," May 24, 1993, p. 1A. *Who said so first?* Bureau of Labor Statistics.

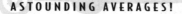

ASTOUNDING AVERAGES!

R & R

Match the country with the average number of vacation days taken annually by its workers.

1. Germany
2. Sweden
3. Britain
4. United States
5. Mexico
6. France

A. 6
B. 10
C. 18
D. 22
E. 25
F. 30

Answers: 1.E 2.F 3.D 4.B 5.A 6.C

Who says so? Newsweek, "The Leisure Lag," February 17, 1992, p. 42. *Who said so first?* Hewitt Associates.

215

POP QUIZ

Gone Fishin'

*T*wenty-five percent of adult Americans—or 46.6 million total—fish for sport; 10 percent—or 16.7 million total—hunt for sport.

Who says so? Council on Environmental Quality, *Environmental Trends.* Washington, D.C.: Council on Envrionmental Quality, 1989, pp. 106–107.

ASTOUNDING AVERAGES!

Take a Hike!

*a*n average of 175 hikers—out of 1,900 who attempt it travel the full Appalachian Trail each year. The average time needed to accomplish this feat is 175 days. To navigate the full length of the trail—2,143 miles, from Maine to Georgia—an average of 5 million steps are needed. An average pair of hiking boots lasts 1,200 to 1,600 miles.

Who says so? *U.S. News & World Report,* "Database," March 23, 1992, p. 14.

That's the Ticket!

n 1940 the average movie ticket cost 24 cents. In 1990 it was $4.75.

Who says so? *The Universal Almanac 1992.* Kansas City, MO: Andrews and McMeel, 1991. p. 253.

218

ASTOUNDING AVERAGES!

Let's Wait 'til It Comes Out on Video

*a*ccording to the Motion Picture Association of America, moviegoers saw an average of 6.7 movies in 1991, down from 7.1 in 1990.

Who says so? American Demographics, "VCR Wars," August 1992, p. 6.

Please Rewind

*I*n 1992, in a population where 75 percent of all households had a VCR (compared with only 5.7 percent in 1982), the average number of movie videos rented per household was 38.

Who says so? *U.S. News & World Report,* "The Balance Sheet, Please . . . ," April 5, 1993, p. 16.

Book Report

*a*ccording to a Gallup Poll, women who read read an average of 18 books per year, but men readers read just 12 books per year. The number of adults who read books today is double the number that read books in the 1950s.

Who says so? Judith Waldrop, "Business Reports—Media: The 1990s Will Be Better for Reading," *American Demographics,* November 1991, p. 17.

221

THAT'S ENTERTAINMENT

Punt, Pass, and Pay

*a*ccording to the newsletter *Team Marketing Report,* in 1992 the average cost for four people to attend a National Football League game—including tickets, concessions, two caps, two programs, and parking—was $163.70.

Who says so? Michael Hiestand, "Average Ticket Prices Up, but 12 Teams Hold Line," *USA TODAY,* September 2, 1992, p. 3C.

ASTOUNDING AVERAGES!

Take Me Out to the Ball Game

*a*ccording to the newsletter *Team Marketing Report*, in 1993 the average cost for four people to attend a major league baseball game was $90.81, an increase of 4.7 percent over 1992's average.

Who says so? *The Wall Street Journal*, "A Special Background Report on Trends in Industry and Finance," April 1, 1993, p. A1.

223

THAT'S ENTERTAINMENT

Maybe They Should Call It Toykyo

*a*ccording to a study conducted by Product Science Research Institute, each child in Tokyo owns an average of 411 toys, totaling $1,530 per child.

Who says so? Detroit Free Press, "Toy Boxes Full in Tokyo," January 28, 1993, p. 1E.

I'm Going to Disney World!

*a*n average London to Orlando package deal to Disney World for plane, hotel, and park entry is $115 per day; but it costs the British $200 per day for the same kind of package deal to Euro Disney. Even without the lodging or travel expenses, a family of four spends an average of $280 per day at Euro Disney, according to a Parisian newspaper.

Who says so? *Business Week*, "The Mouse Isn't Roaring," August 24, 1992, p. 38.

THAT'S ENTERTAINMENT

No Wonder the Games Take So Long

*a*ccording to one estimate, there are 10^{120} possible moves in the average game of chess. A computer would need 10^{101} years to consider all of the moves and find the best response.

Who says so? Elisabeth Geake, "Playing to Win," *New Scientist*, September 19, 1992, p. 24.

Waste Not,
Want Not

Peanuts! Get Your Peanuts!

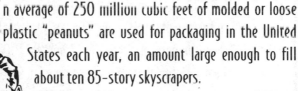

*a*n average of 250 million cubic feet of molded or loose plastic "peanuts" are used for packaging in the United States each year, an amount large enough to fill about ten 85-story skyscrapers.

Who says so? Consumer Reports, "Tis the Season to Recycle," December 1992, p. 745.

It's a Steel

*a*n average of 9,000 steel cans are recovered for recycling every minute of every day, from the 100 million used every day by Americans. Recycling these cans can save 74 percent of the energy needed to produce them from virgin materials.

Who says so? Debi Kimball, *Recycling in America*. Denver, CO: ABC-CLIO, 1992, p. 38.

Clear As Glass

*a*recycled glass container can save the amount of energy needed to light a 100-watt bulb for 4 hours. One ton of recycled glass saves the energy equivalent of 9 gallons of fuel oil. Thirty percent of the average glass soft drink bottle is made up of recycled glass.

Who says so? Debi Kimball, *Recycling in America*. Denver, CO: ABC-CLIO, 1992, p. 39.

It May Not Make Dollars, but It Sure Makes Sense

*a*ccording to Waste Management, a commercial waste hauler, the cost of collecting and sorting an average ton of recyclables is $175. Afterwards, the same ton of recyclables is worth about $44.

Who says so? Jeff Bailey, "Two Major Garbage Rivals Find Their Profits Trashed," *The Wall Street Journal,* March 4, 1993, p. B4.

ASTOUNDING AVERAGES!

Kick the Cans

Every year Americans throw away enough aluminum cans to produce 6,000 DC-10s; enough plastic and paper cups and plates to give the world 6 picnics; and an average of 180 million razors. In addition, each year the British dispose of 2.5 billion diapers and the Japanese dump 30 million single-use cameras.

Who says so? Alan Duening, "How Much is Enough?" *Technology Review,* May–June 1991, p. 60.

Talking Trash

mericans produce an average of 148.1 million tons of garbage a year, or an average of 3.43 pounds of garbage per person per day, up from 2.5 pounds in 1960. If this rate goes unchecked, the average American would produce over 90,000 pounds of garbage in his or her lifetime.

Who says so? Louis Rukeyser, *Louis Rukeyser's Business Almanac.* New York: Simon & Schuster, 1991, p. 28.

ASTOUNDING AVERAGES!

Trash Day

Match the city with the average amount of waste generated daily per citizen.

1. Chicago
2. New York City
3. Rome
4. Los Angeles
5. Paris

A. 6.4 pounds
B. 5.0 pounds
C. 4.0 pounds
D. 2.4 pounds
E. 1.5 pounds

(Answers: 1B, 2C, 3E, 4A, 5D)

Who says so? The 1992 Information Please Environmental Almanac. Boston, MA: Houghton Mifflin Company, 1992, p. 108.
Who said so first? National Solid Wastes Management Association.

Pop Quiz

235

Pounding It Out

*I*n the United States the average waste generated per person per year includes 190 pounds of plastic, 85 pounds of glass, 72 pounds of steel cans, and 24 pounds of plastic containers.

Who says so? Debi Kimball, *Recycling in America*. Denver, CO: ABC-CLIO, 1992, pp. 38–42.

236

All in the Family

a typical suburban family of three generates 40 pounds of garbage weekly. The majority of this garbage consists of yard waste and grass clippings (23 percent), paper (21 percent), and food (11 percent). Miscellaneous materials—such as glass, plastic, metal cans, and aluminum—make up the remaining 45 percent.

Who says so? Good Housekeeping, September 1989, p. 272.

WASTE NOT, WANT NOT

Garbage by the Yard

*a*n average of 3 tons of grass clippings are generated every year by mowing a ½-acre lawn. This is enough to fill 465 one-bushel bags. Twenty to 50 percent of the landfills in the United States are made up of yard waste.

Who says so? Consumer Reports, "The Green Way to a Green Yard," June 1991, p. 407.

You Deserve Some Waste Today

Based on a 1990 week-long waste audit of two McDonald's restaurants—in Sycamore, Illinois, and Denver, Colorado— each day the average McDonald's restaurant serves about 2,000 customers and produces 238 pounds of waste.

Who says so? The Wall Street Journal, April 17, 1991, p. B1. *Who said so first?* McDonald's and the Environmental Defense Fund.

239

WASTE NOT, WANT NOT

Paper Covers Rock

*I*n the United States, every person uses an average of 600 pounds of paper products each year, 70 percent of which ends up in landfills. An average of 40 percent of waste in landfills is paper.

Who says so? *Popular Science,* "Paper Recycling: Fact from Fiction," October 1991, pp. 13–14.

ASTOUNDING AVERAGES!

Extra! Extra! Read All About It!

*a*n average newspaper subscription, such as the *San Francisco Chronicle*, received every day, produces 550 pounds of wastepaper per subscription per year; an average *New York Times* Sunday edition produces 8 million pounds of wastepaper.

Who says so? Audubon, March 1990, p. 4.

241

WASTE NOT, WANT NOT

Save the Planet (While at Work)

*E*very day the average office worker generates an average of 0.51 pounds of recyclable paper. One ton of recycled wastepaper would save an average of 7,000 gallons of water, 3.3 cubic yards of landfill space, 3 barrels of oil, 17 trees, and 4,000 kilowatt-hours of electricity, or enough energy to power the average home for 6 months.

Who says so? Jonathan E. Rinde, "The 'Greening' of Law Offices," *Trial*, May 1991, p. 119.

ASTOUNDING AVERAGES!

The Paper Chase

very year the average office worker generates 180 pounds of high-grade paper. Every year the average 100-person office generates 378,000 sheets of copier paper, or enough paper, if stacked, to equal 7 stories. Every year Americans throw away enough office and writing paper to build a 12-foot wall reaching from Los Angeles to New York City.

Who says so? Debi Kimball, *Recycling in America.* Denver, CO: ABC-CLIO, 1992, p. 50.

Do Not Dispose of in Trash

ifteen pounds of hazardous waste are generated in the average home each year. The majority of this (36.6 percent) is made up of household maintenance items, such as paint, thinners, and adhesives. Other major contributors are household batteries (18.6 percent), personal care products (12.1 percent), cleaners (11.5 percent), and automotive maintenance products (10.5 percent).

Who says so? *Garbage,* March/April 1990, p. 16.

ASTOUNDING AVERAGES!

If They're So Smart, Why Don't They Get Rid of Themselves?

*a*n average of 10 million old mainframes, PCs, and work stations are discarded by American businesses and individuals each year. According to one study, if computers continue to be discarded at this rate, landfill space equal to an acre of land 3 ½ miles deep will be needed to house them by the year 2005. This space is large enough to stack 15 Empire State Buildings end to end.

Who says so? The New York Times, "Recycling Answer Sought for Computer Junk," April 14, 1993, p. A1.

WASTE NOT, WANT NOT

Hold Your Nose
(and Put on Some Sunblock)

*F*or every pound of meat it yields, each head of beef cattle emits about $\frac{1}{3}$ of a pound of methane. Add that to the carbon released from fuels burned in animal farming, and every pound of beef produced has the same greenhouse-warming effect as a 25-mile drive in an average American automobile.

Who says so? Alan Thein Durning, "Fat of the Land: We Can't Keep Eating the Way We Do," *USA TODAY* (magazine), November 1992, p. 27.

ASTOUNDING AVERAGES!

Every 3,000 Miles

*a*n average of 200 to 400 million gallons of waste oil are generated annually by car owners who change their own oil, or enough to equal 36 Exxon *Valdez* oil spills per year. Since only 10 to 14 percent of this oil is disposed of properly, most of it ends up in the ground, in streams, and in sewers. One gallon of used oil can make a million gallons of fresh water undrinkable.

Who says so? Consumer Reports, "Trickle Down," April 1991, pp. 210–211.

Emissions Impossible

or every gallon of gasoline burned, 19 pounds of carbon dioxide are released.

Who says so? *The New York Times,* "A Traffic Jam of Potential Polluters," November 24, 1992, p. C1.

ASTOUNDING AVERAGES!

A Matter of Life
and Death

Must Be Those Cold December Nights

*a*ccording to the National Center for Health Statistics, September is the month with the highest birth rate. In 1990 there were an average of 12,100 babies born every day in September, compared to a daily average of 11,400 for the entire year. Tuesday is the biggest day for births.

Who says so? Judith Waldrop, "Seasons: The Birthday Boost," *American Demographics,* September 1991, p. 4.

Baby Showers

*D*uring the years 1946 to 1964, an average of 4 million babies were born each year, or 76 million total; between 1965 and 1976, an average of 3.4 million babies were born annually, or a total of 41 million babies; since 1977, an average of 3.7 million babies have been born each year.

Who says so? Susan Mitchell, "How to Talk to Young Adults," *American Demographics*, April 1993, p. 50.

Population Explosion

*W*orldwide, an average of 9 babies are born and 3 people die every 2 seconds. The population grows by an average of 93 million people per year, 7.7 million people per month, 1.8 million people per week, 254,000 people per day, 10,600 people per hour, and 3 people every second.

Who says so? The Universal Almanac 1992. Kansas City, MO: Andrews and McMeel, 1991, p. 316. *Who said so first?* U.S. Bureau of the Census, *Worldwide Population Profile: 1989.*

A MATTER OF LIFE AND DEATH

Life Goes On

In the United States the average life expectancy of a child born in 1991 was 75.7 years. The average life expectancy of a child born in the year 2000 is expected to be 77.0 years.

Who says so? U.S. Bureau of the Census, *Statistical Abstract of the United States 1992*. 112th edition. Washington, D.C.: U.S. Department of Commerce, 1992, pp. 824–825. *Who said so first?* U.S. Bureau of the Census, *World Population Profiles: 1991*.

ASTOUNDING AVERAGES!

The Right Stuff

*a*ccording to a study published in the *New England Journal of Medicine*, right-handers live longer than left-handers: an average of 75 years for those who are right-handed, compared to 66 years for left-handed people. Left-handers are six times more likely to die in accidents.

Who says so? *Newsweek*, "Lifestyle—Health: The Right Stuff for a Longer Life," April 15, 1991, p. 58.

255

Bug Off!

*T*he average life expectancy for selected species of insects: fly, 19 to 30 days; firefly, a few weeks; cockroach, 40 days; mosquito, 2 months; cricket, 9 to 14 weeks; grasshopper, 5 months; spider, 4 to 7 years; worker termite, 20 years; and queen termite, 50 years.

Who says so? Frank Kendig and Richard Hutton. *Life-Spans.* New York: Holt, Rinehart and Winston, 1979, pp. 71–76.

ASTOUNDING AVERAGES!

That's 84 Years to You and Me

The average life expectancy for dogs is 12 years. For cats, the average life span of an unaltered male living in a household is 13 to 15 years, while an unaltered female living in a household can expect to live 15 to 17 years. Neutering adds an average of 1 to 2 more years.

Who says so? Guinness Book of Records 1992. New York: Bantam Books, 1992, pp. 70, 74.

For the Birds

*T*he average life expectancy for selected species of birds: woodpecker and mockingbird, 10 years; hawk, 15 years; sparrow, 20 years; ostrich and goose, 25 years; cardinal, 30 years; penguin, 34 years; stork, 35 years; parrot, 50 years; and swan, 70 years.

Who says so? Frank Kendig and Richard Hutton, *Life-Spans.* New York: Holt, Rinehart and Winston, 1979, pp. 52–58.

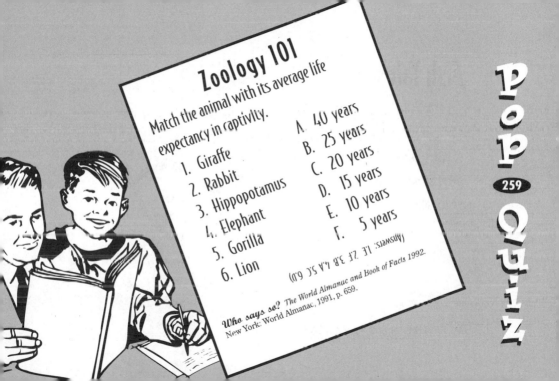

Zoology 101

Match the animal with its average life expectancy in captivity.

1. Giraffe
2. Rabbit
3. Hippopotamus
4. Elephant
5. Gorilla
6. Lion

A. 40 years
B. 25 years
C. 20 years
D. 15 years
E. 10 years
F. 5 years

Answers: 1.E 2.F 3.B 4.A 5.C 6.D

Who says so? The World Almanac and Book of Facts 1992. New York: World Almanac, 1991, p. 659.

Fish Tales

*T*he average life expectancy for selected species of fish: dolphin, 3 to 4 years; white bass, 4 to 5 years; perch, 5 to 6 years; walleye, 6 to 7 years; goldfish, 7 years; pike, salmon, and herring, 10 years; cod, 15 years; and carp, 20 to 25 years.

Who says so? Frank Kendig and Richard Hutton. *Life-Spans.* New York: Holt, Rinehart and Winston, 1979, pp.61–67.

You Can Count the Rings

he average life expectancy for selected species of trees: southern poplar, 100 years; dogwood, 115 years; American elm, 235 years; sugar maple, 275 years; California cedar, 350 years; white oak, 450 years; Douglas fir, 750 years; redwood, 1,000 years; sequoia, 2,500 years; and bristlecone pine, 3,000-plus years.

Who says so? Frank Kendig and Richard Hutton. *Life-Spans.* New York: Holt, Rinehart and Winston, 1979, pp. 107–108.

Planned Obsolescence?

he median life expectancy of an automobile in 1990 was 6.5 years. In 1946 it was 8.8 years.

Who says so? Motor Vehicle Manufacturing Association, *MVMA Motor Vehicle Facts and Figures '91*. Detroit, MI: Motor Vehicle Manufacturing Association, 1991, p. 26.

ASTOUNDING AVERAGES!

And They Always Break at the Worst Possible Time

The average life span of various household appliances: dishwasher, 10 years; air conditioner, microwave oven, and gas water heater, 11 years; gas dryer and automatic washer, 13 years; electric dryer, 14 years; freezer, 16 years; refrigerator and electric stove, 17 years; and gas furnace and gas stove, 19 years.

Who says so? The New York Times, "Appliances Life Spans," December 1, 1990, p. 50.

Let There Be Light

*T*he average life span of a standard 75-watt incandescent light bulb is 750 hours; a long-life 75-watt incandescent bulb, 3,000 hours; a halogen bulb, 3,000 hours; and a compact fluorescent bulb, 10,000 hours.

Who says so? Consumer Reports, "Bright Ideas in Light Bulbs," October 1992, pp. 664–670.

ASTOUNDING AVERAGES!

But Try Keeping It That Long

The average life expectancy of a U.S. one-dollar bill is 1.5 years; a five dollar bill, 2 years; a ten-dollar bill, 3 years; a twenty-dollar bill, 4 years; and a fifty- or one-hundred-dollar bill, 9 years.

Who says so? Los Angeles Public Library/State of California, *Scan/Info*, January 1992, p. 7.
Who said so first? *San Francisco Chronicle*, October 14, 1991, p. D4; and the U.S. Treasury Department.

Accidents Will Happen

*I*n the United States an accidental death occurs every 5 minutes and an accidental injury occurs every 4 seconds. This translates into 1,690 accidental deaths and 165,400 accidental injuries per week.

Who says so? National Safety Council, *Accident Facts 1992*. Itasca, IL: National Safety Council, Safety & Health News Center, 1992, p. 25.

A Goat Might Be Safer

*a*ccording to the Consumer Product Safety Commission, each year in the United States there are an average of 75 deaths due to lawn mower accidents, most often involving persons over the age of sixty-five or under the age of five. Every year there are around 55,000 injuries involving lawn mowers.

Who says so? University of California at Berkeley, School of Public Health, *Wellness Letter*, July 1993, p. 1.

267

No Thanks, I'll Use a Towel

Since January of 1991, hair dryers have been made with a safety plug that automatically shuts the unit off if it falls into water. But plugged-in dryers made before that time are lethal if they land in water. An average of 17 people a year, most of them children under ten, died in this fashion before hair dryers became safer.

Who says so? Consumer Reports, "In the Stores," August 1992, p. 532.

The Sting

*a*n average of 18 to 20 people who are very allergic die from bee stings every year in the United States.

Who says so? Rif S. El-Makkakh and others, "Abuzz Over Bee Keeping," *Science Teacher*, September 1991, pp. 26–31.

A MATTER OF LIFE AND DEATH

The Flu Bug Gets Nasty

*T*here are an average of 20,000 deaths per year due to flu-related complications.

Who says so? U.S. News & World Report, "Database," January 20, 1992, p. 14.

Dying for a Smoke

*W*orldwide, almost 3 million people—including an estimated 400,000 Americans—die annually from smoking-related causes, or an average of 1 death every 10 seconds.

Who says so? Nikolai Khaltaev, "Inter-Health Fights Life-Style Diseases," *World Health*, May–June 1991, pp. 18–20.

They Hardly Had a Chance

*I*n the United States, 9 babies per 1,000 born died before their first birthday in 1991; in China, 22 per 1,000; in Russia, 23 per 1,000; in Mexico, 30 per 1,000; and in Brazil, 55 per 1,000.

Who says so? The New York Times, "Infant Mortality: How Brazil Compares," May 14, 1993, p. A8. *Who said so first?* UNICEF and World Bank Atlas.

Back to Nature

Heads Up!

*a*n average of 20 million tons of particles of varying size fall to the earth each day, some no bigger than a grain of sand.

Who says so? Sharon Begley, "The Science of Doom," *Newsweek*, November 23, 1992.

The Calm Before the Storm

When a hurricane is forecast to move inland on a projected path, the coastal area placed under warning is about 300 miles in length. The average cost of preparations, whether a hurricane strikes or not, is more than $50 million for the Gulf Coast. This estimate covers the cost of boarding up homes, closing down businesses and manufacturing plants, etc.

Who says so? Frank E. Bair, *The Weather Almanac.* 6th edition. Detroit, MI: Gale Research, 1992, p. 59.

ASTOUNDING AVERAGES!

Everyone Knows It's Windy

Hurricane winds are divided into three regions. The outer portion of hurricane winds blow with an average force of 40 miles per hour; the second region of winds blows with an average force of 74 miles per hour; and the third region is of maximum winds, which have an average speed of 120 to 150 miles per hour.

Who says so? Barbara Tufty, *1001 Questions Answered About Hurricanes, Tornadoes and Other Natural Air Disasters*. New York: Dover Publications, 1987, p. 22.

Seems Like a Lifetime

*H*urricanes of average size take about 10 hours to pass a given spot. The first part of the storm lasts an average of 4 hours; the eye, or quiet time, lasts an average of 20 to 30 minutes; and the final part of the hurricane lasts another 4 to 5 hours before the winds calm down and the rains cease.

Who says so? Barbara Tufty, *1001 Questions Answered About Hurricanes, Tornadoes and Other Natural Air Disasters.* New York: Dover Publications, 1987, p. 24.

ASTOUNDING AVERAGES!

Now All We Have to Do Is Harness It

he average heat energy released by a hurricane in one day can be the equivalent of energy released by fusion of 400 20-megaton hydrogen bombs. This released energy, if converted to electricity, could supply the electrical needs of the entire United States for about 6 months.

Who says so? Frank E. Bair, *The Weather Almanac.* 6th edition. Detroit, MI: Gale Research 1992, p. 51.

BACK TO NATURE

Shakin' All Over

*a*n average of 2 earthquakes shake the earth every minute, or about 1 million earthquakes per year. An average of 20 major earthquakes, measuring 7 to 7.9 on the Richter Scale, break out every year. Great earthquakes, measuring 8 or above on the Richter Scale, occur on average once every 2 to 3 years.

Who says so? Barbara Tufty, *1001 Questions About Earthquakes, Avalanches, Floods and Other Natural Disasters.* New York: Dover Publications, 1978, p. 18.

ASTOUNDING AVERAGES!

On Shaky Ground

entral Nevada is the most active earthquake region in the United States, having been shaken by an average of 5,000 earthquakes per year since 1769, the date when records were first kept.

Who says so? Barbara Tufty, *1001 Questions About Earthquakes, Avalanches, Floods and Other Natural Disasters.* New York: Dover Publications, 1978, p. 35.

BACK TO NATURE

Small but Deadly

n average, tornado paths are only ¼ of a mile wide and seldom more than 16 miles long.

Who says so? Frank E. Bair, *The Weather Almanac.* 6th edition. Detroit, MI: Gale Research, 1992, p. 84.

40 MPH in a 25 Zone

*T*he forward speed of a tornado averages 40 miles per hour.

Who says so? Frank E. Bair, *The Weather Almanac*. 6th edition. Detroit, MI: Gale Research, 1992, p. 84.

283

Raindrops Keep Fallin' on My Head

*T*he speed of a raindrop varies with drop size and wind speed. A typical raindrop falls an average of 600 feet per minute or 7 miles per hour.

Who says so? Frank H. Forrester, *1001 Questions Answered About the Weather.* New York: Grosset & Dunlap, 1957, p. 53.

Thunder Only Happens When It's Raining

*a*n average of 2,000 thunderstorms are forming, growing, or dying somewhere on earth every minute. An average of 16 million such storms are formed each year.

Who says so? Barbara Tufty, *1001 Questions Answered About Hurricanes, Tornadoes and Other Natural Air Disasters.* New York: Dover Publications, 1987, p. 114.

Rain Delay

*T*he average length of a thunderstorm is 25 minutes. It drops an average of 500,000 tons of water, or about ¾ of an inch of rain over a 9-square-mile area.

Who says so? Frank Field, *Doctor Frank Field's Weather Book*. New York: Putnam, 1981, p. 196.

ASTOUNDING AVERAGES!

No Wonder They Go Boom

ccording to some scientists, an average thunderstorm, 25,000 to 40,000 feet high, releases more energy per minute than a 120-kiloton nuclear bomb.

Who says so? Barbara Tufty, *1001 Questions Answered About Hurricanes, Tornadoes and Other Natural Air Disasters.* New York: Dover Publications, 1987, p. 114.

And Never the Same Place Twice

t is estimated that lightning strikes the earth an average of 100 times each second.

Who says so? Frank E. Bair, *The Weather Almanac.* 6th edition. Detroit, MI: Gale Research, 1992, p. 120.

Don't Forget Your Umbrella

Tutunendo, Colombia, has the greatest annual rainfall In the world, an average of 463.4 inches per year.

Who says so? *Guinness Book of Records 1992.* New York: Bantam Books, 1992, p. 44.

It's a Marvelous Night for a Rain Dance

*a*n annual average of 0.02 inches of rain falls at Arica, on the northern desert of Chile. Many years may pass with no rain at all.

Who says so? Barbara Tufty, *1001 Questions About Earthquakes, Avalanches, Floods and Other Natural Disasters.* New York: Dover Publications, 1978, p. 254.

Brrrrrrrrrr!

*T*he lowest average annual temperature in the United States is 9.6 degrees Fahrenheit at Barrow, Alaska. Barrow also has the coolest summers, with an average temperature of 36.7 degrees Fahrenheit. The lowest average winter temperature is -15.6 degrees Fahrenheit at Barter Island on the Arctic coast of Alaska.

Who says so? Frank E. Bair, *The Weather Almanac.* 6th edition. Detroit, MI: Gale Research, 1992, p. 358.

BACK TO NATURE

Hot Fun in the Summertime

he highest average annual temperature in the United States is 77.7 degrees Fahrenheit at Key West, Florida. The highest average summer temperature is 98.2 degrees Fahrenheit at Death Valley, California. The

highest average winter temperature is 70.2 degrees Fahrenheit at Key West.

Who says so? Frank E. Bair, *The Weather Almanac.* 6th edition. Detroit, MI: Gale Research, 1992, p. 357.

ASTOUNDING AVERAGES!

Sunny Today (or not)

Match the city with its average percentage of possible sunshine.

1. Charleston, WV
2. Denver, CO
3. Juneau, AK
4. Sacramento, CA
5. Phoenix, AZ
6. Cleveland, OH

A. 86%
B. 78%
C. 70%
D. 49%
E. 40%
F. 30%

(Answers: 1.E 2.C 3.F 4.B 5.A 6.D)

Who says so? U.S. National Oceanic and Atmospheric Administration, *Comparative Climatic Data,* annual.

It's Not the Heat, It's the Humidity

Cincinnati, Ohio, is the most humid city in the United States, with an average relative humidity of 81 percent; Phoenix, Arizona, is the least humid U.S. city, with an average relative humidity of 23 percent.

Who says so? U.S. Bureau of the Census, *Statistical Abstract of the United States, 1992.* 112th edition. Washington, D.C.: U.S. Department of Commerce, 1992, p. 227. *Who said so first?* U.S. National Oceanic and Atmospheric Administration, *Comparative Climatic Data,* annual.

Pssst! Don't Tell Chicago!

For a period of record through 1990, Cheyenne, Wyoming, topped the list of the windiest cities in the United States with an average wind speed of 13.0 miles per hour. Chicago ranked twenty-first on the list with an average wind speed of 10.3 miles per hour.

Who says so? U.S. Bureau of the Census, *Statistical Abstract of the United States 1992.* 112th edition. Washington, D.C.: U.S. Department of Commerce, 1992, p. 227. *Who said so first?* U.S. National Oceanic and Atmospheric Administration, *Comparative Climatic Data,* annual.

295

BACK TO NATURE

Ice Fishing Darn Near Impossible

he average thickness of the ice that covers Antarctica is 7,100 feet. However, at its thickest point it is 15,700 feet. This is 10 times taller than the Sears Tower in Chicago, the world's tallest building.

Who says so? World Book Encyclopedia. Chicago, IL: World Book, 1990, vol. 1, pp. 530, 532.

Breaking the Ice

*a*n average size iceberg would need an estimated 2,000 tons of TNT to break it up or 2 million gallons of gasoline to melt it down.

Who says so? Barbara Tufty, *1001 Questions Answered About Hurricanes, Tornadoes and Other Natural Air Disasters.* New York: Dover Publications, 1987, p. 305.

Off the Deep End

*T*he average depth of the Pacific Ocean is 13,740 feet. It is the world's largest ocean, measuring 64.2 million square miles in area and accounting for 45.9 percent of the world's oceans.

Who says so? Guinness Book of Records 1992. New York Bantam Books, 1992, p.18..

Take Our Word for It

*T*he average porcupine has about 30,000 quills.

Who says so? David F. Costello, *The World of the Porcupine*. Philadelphia, PA: Lippincott, 1966, p. 13.

299

BACK TO NATURE

Charlotte's Web

he average orb-weaver spider takes 30 to 60 minutes to completely spin its web. The web must be replaced every few days because it loses its stickiness, and thus its ability to entrap food.

Who says so? *The Illustrated Encyclopedia of Wildlife.* Lakeville, CT: Grey Castle Press, 1991, vol. 9, pp. 2264–2265.

You Can Watch It Grow

Bamboo plants grow an average of 35.4 inches every day.

Who says so? Diagram Group, *Comparisons*. New York: St. Martin's Press, 1980, p. 186.

Index

A

Accidental death *266*
Advertising *137–138, 209–211, 213*
Age *112–114*
Airfare *196*
Airplane trips *197*
Airports *195*
Appliances *263*
Automobile accidents *191*

B

Babies *252, 272*
Baby food *46*
Bamboo plants *301*
Barbie dolls *88*
Baseball games *223*
Bee stings *269*
Birds *258*
Birth rate *251*
Birth weight *108*
Birthday gifts *107*
Birthdays *107*
Blinking *158*
Bones *156*
Books *221*
Brains *169*
Business travelers *76*

C

Caffeine *47*

ASTOUNDING AVERAGES!

ASTOUNDING AVERAGES!

307

ASTOUNDING AVERAGES!

309

310

ASTOUNDING AVERAGES!